Cover

The cover picture shows mouse parietal mesoderm viewed by laser scanning confocal microscopy. The image was obtained 24 h after intraperitoneal injection of crocidolite asbestos fibres. An asbestos fibre (pseudo-colour green) is located within a proliferating mesothelial cell, which is visualized by immunofluorescence detection of nuclear-incorporated bromodeoxyuridine (pseudocolour red). Tissue preparation and photography by Janice Macdonald and Agnes Kane, Brown University, Providence, Rhode Island, USA.

International Agency For Research On Cancer

The International Agency for Research on Cancer (IARC) was established in 1965 by the World Health Assembly as an independently financed organization within the framework of the World Health Organization. The headquarters of the Agency are in Lyon, France.

The Agency conducts a programme of research concentrating particularly on the epidemiology of cancer and the study of potential carcinogens in the human environment. Its field studies are supplemented by biological and chemical research carried out in the Agency's laboratories in Lyon and, through collaborative research agreements, in national research institutions in many countries. The Agency also conducts a programme for the education and training of personnel for cancer research.

The publications of the Agency contribute to the dissemination of authoritative information on different aspects of cancer research. A complete list is printed at the back of this book. Information about IARC publications, and how to order them, is also available via the Internet at: **http://www.iarc.fr/**

Mechanisms of Fibre Carcinogenesis

Edited by A.B. Kane, P. Boffetta, R. Saracci and J.D. Wilbourn

IARC Scientific Publications No. 140

International Agency for Research on Cancer, Lyon, 1996

Published by the International Agency for Research on Cancer,
150 cours Albert Thomas, F-69372 Lyon cedex 08, France

© International Agency for Research on Cancer, 1996

Distributed by Oxford University Press, Walton Street, Oxford, UK OX2 6DP (Fax: +44 1865 267782) and in
the USA by Oxford University Press, 2001 Evans Road, Carey, NC 27513, USA (Fax: +1 919 677 1303).
All IARC publications can also be ordered directly from IARC*Press*
(Fax: +33 4 72 73 83 02; E-mail: press@iarc.fr).

Publications of the World Health Organization enjoy copyright protection in
accordance with the provisions of Protocol 2 of the Universal Copyright Convention.
All rights reserved.

The designations used and the presentation of the material in this publication do not imply the
expression of any opinion whatsoever on the part of the Secretariat of the World Health Organization
concerning the legal status of any country, territory, city, or area or of its authorities,
or concerning the delimitation of its frontiers or boundaries.

The mention of specific companies or of certain manufacturers' products does not imply
that they are endorsed or recommended by the World Health Organization in preference to others
of a similar nature that are not mentioned. Errors and omissions excepted,
the names of proprietary products are distinguished by initial capital letters.

The authors alone are responsible for the views expressed in this publication.

The International Agency for Research on Cancer welcomes requests for permission to
reproduce or translate its publications, in part or in full. Applications and enquiries should be addressed
to the Editorial & Publications Service, International Agency for Research on Cancer,
which will be glad to provide the latest information on any changes made to the text, plans for new editions,
and reprints and translations already available.

IARC Library Cataloguing in Publication Data

Mechanisms of fibre carcinogenesis/
editors, A.B. Kane...(et al.)
(IARC Scientific Publication; 140)

1. Carcinogens – toxicity
2. Minerals – toxicity
3. Neoplasms – etiology

I. Kane, A.B. II. Title III. Series

ISBN 92 832 2140 0   (NLM Classification: W1)
ISSN 0300–5085

Foreword

While several types of mineral fibres, such as asbestos and erionite, are known to pose a carcinogenic hazard to humans, the evidence regarding other types of fibres, such as ceramic fibres and glasswool, is a matter of debate. In the latter case, the integration of results from human and experimental studies is often a critical step in the process of assessing the risk to humans, and an understanding of the mechanisms through which mineral fibres exert their carcinogenic activity is of major importance in this process.

This publication originates from a Workshop held at IARC in Lyon on 9–11 January 1996, which was attended by invited experts from different countries. The workshop had two goals: to review and discuss the current knowledge on mechanisms of fibre carcinogenicity, and to formulate recommendations to IARC on the use of such data in the process of evaluation of carcinogenic risks to humans within the framework of the IARC Monographs programme. The main outcome of the workshop was the Consensus Report, which is published in the first part of this volume, and which was agreed by the whole group. In addition, selected participants in the workshop were invited to prepare and present authored papers on specific aspects of mineral fibre carcinogenicity. These contributions, which report the opinions of their authors, are also included in this volume. Some of these papers address controversial issues on which no complete agreement was reached within the Group, and perhaps even less agreement exists in the scientific community at large.

The workshop was partially supported by the US National Institute for Environmental Health Sciences. The dedicated work of all invited experts, and in particular of Dr Agnes Kane, who chaired the meeting, is greatly appreciated.

P. Kleihues
Director, IARC

List of participants

Meeting on the mechanisms of fibre carcinogenesis
Lyon, 9–11 January 1996

J.C. Barrett[a]
Program of Environmental
Carcinogenesis
National Institute of
Environmental Health
Sciences
PO Box 12233
Research Triangle Park
NC 27709, USA

D.M. Bernstein
Consultant in Toxicology
40 chemin de la Petite
Boissière
1208 Geneva
Switzerland

J. Bignon
INSERM, Unité 139
Faculté de Médecine
8 rue du Général Sarrail
94010 Créteil Cédex
France

B. Case
Departments of Pathology
and Occupational Health
McGill University
Montreal
Québec H3A 2B4
Canada

C. Collier
Inhalation Toxicology
AEA Technology
551 Harwell
Didcot OX11 0RA
United Kingdom

J. Davis
Institute of Occupational
Medicine
Roxburgh Place
Edinburgh EH8 9S8
United Kingdom

K. Donaldson
Department of Biological
Sciences
Napier University
10 Colinton Road
Edinburgh EH10 5DT
United Kingdom

K. Driscoll
Procter & Gamble Co.
Miami Valley Laboratories
1 Procter & Gamble Plaza
PO Box 538707
Cincinnati, OH 45253-8707
USA

J. Everitt
Chemical Industry Institute
of Toxicology
6 Davis Drive
PO Box 12137
Research Triangle Park
NC 27709
USA

B. Fubini
University of Torino
Dipartimento di Chimica
Inorganica
Chimica Fisica e Chimica
dei Materiali
via Pietro Guiria 9
10125 Torino
Italy

T. Hei
Department of Radiation
Oncology
College of Physicians and
Surgeons of Columbia
University
630 West 168th Street
New York, NY 10032
USA

T.W. Hesterberg
Health, Safety and
Environment Department
Schuller International, Inc.
PO Box 625005
Mountain Technical Center
Littleton, CO 80162-5005
USA

M.-C. Jaurand
INSERM, Unité 139
CHU Henri Mondor
51 avenue Maréchal de
Lattre de Tassigny
94010 Créteil Cédex
France

A.B. Kane (co-Chair)
Department of Pathology
Brown University
Biomedical Center
Box G-B511
Providence
RI 02912
USA

K. Linnainmaa
Institute of Occupational
Health
Topeliuksenkatu 41 a A
00250 Helsinki
Finland

H. Muhlé
Fraunhofer-Institut für
Toxikologie und
Aerosolforschung
Nikolai-Fuchs-Strasse 1
30625 Hannover
Germany

G. Oberdörster
Department of Environmental
Medicine
School of Medicine and
Dentistry
575 Elmwood Avenue
EHSC
Rochester, NY 14642
USA

M. Roller
Medizinische Institut für
Umwelthygiene an der
Heinrich-Heine Universität
Auf'm Hennekamp 50
40225 Düsseldorf
Germany

J.H. Roycroft
Toxicology Branch
Environmental Toxicology
Program
National Institute of
Environmental Health
Sciences
PO Box 12233
Research Triangle Park
NC 27709
USA

R. Saracci (co-Chair)
Istituto Fisiologia Clinica (CNR)
Epidemiology Section
via Trieste
56100 Pisa
Italy

L. Simonato[a]
Centro Oncologico
Regionale
Registro Tumori del Veneto
c/o Direzione Sanitaria
via Giustinian 2
35100 Padova
Italy

[a]Unable to attend.

IARC Participants
P. Boffetta
C. Malaveille
D. McGregor
E. Merler
H. Ohshima
C. Partensky
J. Wilbourn
H. Yamasaki

Technical Editor
W.G. Morgan
94 Bruntsfield Place
Edinburgh EH10 4ES
United Kingdom

Contents

Consensus report	1
Mechanisms of mineral fibre carcinogenesis A.B. Kane	11
Use of physico-chemical and cell-free assays to evaluate the potential carcinogenicity of fibres B. Fubini	35
Use of in-vitro genotoxicity and cell transformation assays to evaluate the potential carcinogenicity of fibres M.-C. Jaurand	55
Effects of fibres on cell proliferation, cell activation and gene expression K.E. Driscoll	73
Short-term animal studies for detecting inflammation, fibrosis and pre-neoplastic changes induced by fibres K. Donaldson	97
Evaluation and use of animal models to assess mechanisms of fibre carcinogenicity G. Oberdörster	107
Mixed fibrous and non-fibrous dust exposures and interactions between agents in fibre carcinogenesis J.M. Davis	127

Consensus report

This Consensus Report was prepared by a group of experts convened at IARC on 9–11 January 1996. The first part of the report addresses the strengths, weaknesses and gaps in the present knowledge on fibre characterization, genotoxicity, cell proliferation and activation, and animal studies. The second part of the report provides answers to specific questions on the relevance of mechanistic data from in-vitro and in-vivo assays in the assessment of the carcinogenic risk of fibres to humans. Finally, the relevance of mechanistic data in the evaluation of fibre carcinogenicity is discussed.

Fibre characterization

Irrespective of study type or design, the full characterization of all particulate material in a test sample is an essential step in the understanding of the mechanisms of fibre carcinogenesis. There is a broad consensus on many of the parameters necessary for fibre characterization and on the methodology needed to obtain these data.

In terms of the form of the test material, the dimensions of all particles (fibrous and non-fibrous) are important, and are best expressed as bivariate size distributions, i.e. diameter (which primarily determines respirability) and length (which is related to biological activity). Methods of fibre characterization range from relatively simple procedures (such as the use of optical or electron microscopy), to highly specialized techniques (such as laser-assisted microprobe mass spectrometry analysis or LAMMA). Aspect ratios (length : diameter) and size-specific fractions have both biological and regulatory significance, and measurement of dimensions is best accomplished by transmission electron microscopy (TEM). Scanning electron microscopy (SEM) can be equally effective in cases where its magnification and contrast are adequate for the fibre type(s) in question. Micromorphology, measurable by TEM but rarely reported for samples used in biological systems, consists of steps, kinks, edges, etc., and contributes to both reactivity and durability. Specific particle surface area may also be measured using low-temperature nitrogen adsorption. Crystallinity can be evaluated by X-ray diffraction or by selected area electron diffraction (SAED). Although crystallinity is of most importance in relation to mineral fibres, some ceramic materials are crystalline and at least one – silicon carbide 'whiskers' – shows biological effects in both animals (cytotoxicity) and humans (fibrosis).

All aspects of the chemical composition of test materials are important, including the intrinsic chemical constituents of fibres, their surface chemistry and the identity of any chemicals derived from contaminants or acquired during storage. One example of chemical composition that has potential mechanistic importance is iron content. Iron is particularly important in the generation of reactive oxygen/nitrogen species (ROS) at the surface of asbestos fibres. In amphiboles (such as amosite and crocidolite), iron is an intrinsic constitutive component, and this may be significant in view of the generally observed greater carcinogenic potential of these fibre types in humans. However, iron may also be present on other fibre types, for example substituting for magnesium ions in chrysotile. This is a highly variable phenomenon that may vary from sample to sample. Iron may also be present in man-made fibres, as a constituent or as an impurity. Which type(s) of iron, if any, have importance for carcinogenicity via ROS-related mechanisms is unknown; there does not appear to be a dose-relationship between iron content and the generation of ROS.

ROS generation at the solid–liquid interface can be measured by spin trapping or by secondary biochemical reactions, such as DNA damage or lipid peroxidation.

Ferrous iron and other ions are exposed in milling, and asbestos minerals may be activated or otherwise altered: ROS release, among other effects, is modified. Metal contamination of fibres has been reported from the milling device.

The importance of biopersistence is well recognized and was the subject of a recent IARC/INSERM/CNRS symposium (see Bignon *et al.*, 1994). Reingestion cycles for macrophages may also be important. Biopersistence studies *in vivo* (animals and humans) and the more recently developed assays of fibre durability in cellular and acellular in-vitro studies should now be a routine part of particle characterization.

Strengths, weaknesses and gaps in fibre characterization

There is a large and historically useful literature on many aspects of fibre characterization. This is a multidisciplinary field based on industrial hygiene, physical and inorganic chemistry, inhalation toxicology, experimental pathology and mineralogy. Much information is available on the measurement of fibre characteristics and there is a considerable data bank extending back for many years. Investigators of new biological phenomena must be aware of this and also be familiar with the fibre characterization techniques described above and in the authored paper by B. Fubini in this volume. New techniques, such as atomic force microscopy, have emerged in the last five years, and investigators should be aware of these.

While it is acknowledged that crystallinity and micromorphology have important effects on fibre reactivity and durability, these parameters are rarely specified in sample characterization for biological studies. There are also few measurements available on specific particle surface area.

Data have been lacking on a number of chemical parameters, including (but not limited to) adsorptive capacity for both exogenous and endogenous materials. There is also a need for the analysis of bulk composition (full chemical analysis) and surface composition (including redox states of ions). Particle chemistry has traditionally been measured by electron diffraction (ED) analysis coupled with TEM or SEM, but these methods give only a proportional chemical make-up, mainly of the particle surface and the subsurface layers. There are some data on exogenous and endogenous adsorbed materials and on the many other chemical parameters that should also be measured. These data are widely available, although interlaboratory comparisons are difficult due to different techniques of sample preparation and analysis.

There remains taxonomic confusion and lack of standardized operating definitions for fibres. 'Asbestos' is often inappropriately used as a generic, homogeneous rubric, and even when an asbestos fibre type is specified, its source is rarely stated. Even standardized samples often contain mixtures of fibres. For example, the 'Union Internationale Contre le Cancer (UICC) B' chrysotile sample is a blend from eight different sources, which are known to have differences in human disease potential.

Terms such as 'asbestiform' and even 'fibre' are highly controversial across disciplines; for example, most mineralogists would not consider particles having an aspect ratio of less than 10 : 1 as 'fibrous'. In this report, however, the WHO definition of a fibre (aspect ratio greater than 3 : 1) is used (WHO/EURO Technical Committee for Monitoring and Evaluating Airborne MMMF, 1985) The definition of what constitutes 'long' or 'short' fibres is also debatable. However, this may not be a problem if bivariate size distributions are obtained and 'biological operational definitions' are used; for example, an important criterion of fibre length may be based on the species-specific ability of macrophages to phagocytize a particular fibre completely. There needs to be agreement as to the best expression of fibre 'exposures' in in-vitro and in-vivo systems. Numerators consisting of fibre numbers or surface area are preferable to fibre mass.

There is more unanimity on the physical characteristics of man-made vitreous fibres (MMVF) and other man-made fibres. A serious question must be asked as to whether 'asbestos' or other natural fibres can be as well characterized for biological studies as man-made fibres – certainly the task is more difficult.

The chemical make-up of most mineral fibre samples has not been adequately characterized. Iron content, for example, can be in the form of an endogenous coating, as in asbestos bodies, and the surface composition may not reflect the constituent chemical composition. Furthermore, the type of iron or other transition elements on or in a fibre that are potentially toxic is also unknown. Quantitation of ROS present in culture media is not usually performed. Whether natural or man-made, inorganic and organic (e.g. cellulose, *para*-aramid) fibres should be fully characterized by dimensions, chemistry and biopersistence when used in any well-conducted biological study. The following information should also be provided: time and place of acquisition (for example, mine coordinates, geological site, workplace sample, etc.); contaminants (mineral and chemical); and storage conditions (time of storage, effects of storage solutions on particle chemistry and possible oxidative changes, possible adsorption from containers). The potential for human exposure to the fibre sample should also be stated.

Fibre characterization should be performed before, during, and at the end of experiments.

Genotoxicity, cell proliferation and cell activation
Carcinogenesis by fibres appears to be a multistage process and may arise by the ability of fibres to cause (i) altered expression or function of key genes arising from genetic or epigenetic alterations; (ii) altered cell proliferation; (iii) altered regulation of apoptosis; or (iv) chronic, persistent inflammation.

In the lung, multiple cell types have been shown to proliferate in response to deposited fibres, including those that are not targets for neoplastic change by fibres. Therefore, additional mutagenic and/or clastogenic events are required for the development of tumours. From in-vitro cell assays, all asbestos fibres tested have demonstrated potential genotoxicity in mammalian cells. From studies with asbestos, it is clear that these fibres can activate macrophages and epithelial cells to release inflammatory mediators, cytokines and growth factors that may alter epithelial and mesothelial cell proliferation and differentiation. Recent data have shown the following: (i) fibres can bind to the plasma membrane and activate cells; and (ii) asbestos fibres can activate multiple intracellular signalling pathways and transcription factors. Oxidative stress may be central to the fibre-mediated activation of key inflammatory/proliferative signalling pathways via redox-sensitive transcription factors such as nuclear factor κB (NFκB) and the activator protein AP1.

Strengths, weaknesses and gaps in genotoxicity studies
Genotoxicity studies have provided mechanistic information at the cellular and molecular level. Cytogenetic studies have suggested that chromosomal aberrations, deletions and aneuploidy are relevant for fibre carcinogenesis. There is a fairly good consensus between these types of studies, which have been carried out with different types of mammalian cells.

The major weakness of these in-vitro genotoxicity studies is lack of validation *in vivo*. In addition, the total data bank on fibres is small and only a few fibre types have been tested. The influence of fibre durability has not yet been assessed in in-vitro genotoxicity assays. In addition, the following shortcomings in experimental design can be identified in many of the in-vitro cell studies: an absence of adequate statistical analysis; a lack of positive and negative controls; and the potentiation of DNA damage caused by either exogenous components (e.g. iron) or cell-derived components (e.g. oxidized lipids) in the culture media.

Limited data are available on the formation of oxidized bases (e.g. 8-hydroxydeoxyguanosine, 8-OHdG) in in-vitro cell models. The specific oxidizing agent (hydroxyl radical, OH•; peroxynitrite; lipid hydroperoxides) or clastogenic factors have not been identified.

Strengths, weaknesses and data gaps in cell proliferation and activation studies
A variety of target cells proliferate following short-term in-vivo exposure to fibrous materials; these observations are consistent with fibre-induced disease outcomes in other studies. Specifically, long crocidolite asbestos fibres induce greater lung epithelial and/or mesothelial cell proliferation in intratracheally and intraperitoneally exposed rats than do short fibres. Similarly, in agreement with tumour outcome in inhalation bioassays, ceramic-fibre-exposed hamsters show greater mesothelial cell proliferation than do rats. Lastly, in rats, mesothelial cell proliferation following ceramic fibre exposure is greater than that induced by glass fibre in apparent correlation with tumour outcome in recent chronic inhalation studies.

Studies using cells exposed to fibres *in vitro* have demonstrated fibre-specific activation, which correlates with the pathological responses observed *in vivo*. For example, cytokine and oxidant release by macrophages *in vitro* is greater in response to long amphibole asbestos than to a short fibre preparation obtained from the same material. In addition, cytokine release from pulmonary epithelial cells was greater in crocidolite-exposed cells than in glass-fibre-exposed cells.

There are multiple weaknesses in the published in-vitro cellular assays. In general, there is a poor characterization of dose with regard to fibre number, dimensions and surface area; also, the specific dose internalized by the target cells is rarely determined. The lung environment modifies the surface of fibres by coating them with endogenous molecules before they make contact with cells; this is seldom addressed in in-vitro studies. Cell lines are highly selected and therefore it is uncertain whether findings obtained with them are representative of target cells *in vivo*. Stable cultures of mesothelial cells are difficult to obtain for in-vitro studies. Most assays are based on the response of the whole cell population; since cells are known to be heterogeneous, methods that allow determination of response in individual cells are desirable. There is limited standardization between laboratories in the assessment of methods and this precludes critical evaluation of differences in the effects reported; for example, there is limited standardization in culture conditions and media composition, including the levels of serum and growth factors. There is a need for in-vivo validation and confirmation of in-vitro findings.

In short-term in-vivo studies, intratracheal instillation and intracavity injection are widely used methods of administration, but the results can be difficult to interpret because of high-dose exposures that may result in uneven deposition. Much of the data bank on the rodent mesothelioma cell and its molecular biology comes from tumours produced in animals exposed by intraperitoneal or intrapleural instillation.

The mechanism or combinations of mechanisms that underlie the epithelial and mesothelial cell proliferative response to fibres are unclear. It is not known whether there are multiple fibre-specific and cell-specific activation pathways or a common mechanism of cell activation. The relationship between cell proliferation in different asbestos-related diseases and in-vivo genotoxic effects is unknown. Likewise, the mechanisms responsible for proliferative responses in short- versus long-term exposures and to low- versus high-dose exposures remain to be elucidated. It is premature to derive generalizations about mechanisms because studies have been limited to a restricted number of fibres. In many cases there has not been a wide enough dose–response relationship demonstrated to allow comparisons between fibre types and preparations to be made.

Fibres may stimulate cell proliferation by several pathways: (i) activation of intracellular signalling pathways mimicking growth factors; (ii) stimulation of growth factor production; and (iii) up-regulation of growth factor receptor expression. Information on defining the receptors that fibres interact with at the cell surface is limited, and the data that are available are confined largely to macrophages exposed to asbestos. There is also limited information on receptors for fibres on epithelial cells and mesothelial cells and for man-made fibres in general.

Qualitative or quantitative differences in the proliferative response of rodent mesothelial cells have been reported, but it is unclear how these correlate with mesothelioma induction. Since a variety of soluble lung toxins can cause mesothelial and epithelial proliferation, there is a gap in the knowledge regarding the mechanism of fibre-specific proliferation. This gap extends to the cytokinetics of epithelial and mesothelial cells and the factors that lead to sustained proliferation in the case of fibres. There is a particular gap in the knowledge concerning the responses of the parietal pleural mesothelium. Limited information is available on the ability of fibres other than asbestos to cause cell activation and proliferation.

In mechanistic terms, little is known of the relationship between fibrosis and neoplasia, although the inflammatory fibroproliferative environment may promote neoplasia because of the localized accumulation of mitogenic mediators from activated inflammatory cells and fibroblasts.

There is a gap in the knowledge of the relative importance of fibre-derived and cell-derived ROS and their interaction in causing oxidative stress. There is a need for data to fill the gap on the adaptive responses of fibre-exposed tissue including induction of antioxidant defences and the impact of fibre-mediated oxidative stress on the redox state of target cells.

In using markers of proto-oncogene and tumour suppressor gene alteration in trying to understand the mechanism of mesothelioma induction, the use of cell lines and mature tumours may be a limitation because important events may have occurred early *in situ*. Time-course studies of the lesions *in situ* are necessary to investigate these early changes.

Animal studies
All animal studies with fibres should be designed and conducted to give exposure–dose–response relationships using more than two exposure–dose levels, with the highest dose being at the maximal tolerated dose (MTD). There is, however, no consensus on how the MTD should be determined in a range of experimental situations. An initial

definition of an MTD that has been used for long-term carcinogenicity studies is 'a dose that produces no increased mortality compared to controls, no shortening of life span other than that resulting from tumour development and no more than a 10% weight gain reduction compared to controls'. This definition, however, is not adequate for all types of studies, and the MTD might have to be defined differently according to the exact processes examined. In studies on fibres, indicators of MTD that might be of widespread usefulness include the following, where applicable: increases in inflammatory parameters; increased target cell proliferation; altered histopathology other than carcinogenicity; prolonged lung clearance function; and the existence of non-linear fibre retention kinetics. In most cases, several of these indicators should be considered. However, the magnitude of change in each case has not yet been determined.

Strengths, weaknesses and data gaps in animal studies
Weaknesses of both short-term and long-term animal studies that can be specified from existing publications include inadequate details of techniques used and most particularly inadequate quantification of the fibres used. The exact details of fibre length and diameter (bivariate analysis) should always be recorded as well as the amount of non-fibrous particulate material in the dust specimens. In addition, most studies published so far lack information on dose–response, a most important aspect.

The major strengths of short-term studies are that, because they are both shorter and less expensive than long-term studies, a wide range of materials can be examined at one time. At present, they are mainly used to examine mechanisms of fibre–cell interactions. These studies should not be limited to short-term work, since one of their main weaknesses at present is that mechanisms are studied for a few days when information on the longevity of the process is needed. Many short-term results are suggested as predictors (biomarkers) of long-term pathological change, but, as yet, there have been few confirmations from long-term studies that these events actually occur. This is one of the major gaps that needs to be filled before we can evaluate fully short-term mechanistic studies.

Long-term studies can be used to demonstrate that mechanistic changes first examined in short-term tests are important throughout actual disease development. At present there is no consensus on the best animal species to mimic and predict human reactions. The rat is most commonly used in chronic inhalation studies but there are data suggesting that the hamster may be better for studies on mesothelioma development. A multidose asbestos inhalation study in rats and hamsters is needed, although, in future, other species may be considered; a well-characterized amphibole asbestos, a known human carcinogen, should be used. Such a study is not available at present. This circumstance was identified as a significant gap. Weaknesses and gaps in studies so far published relate mainly to the general considerations of inadequate fibre characterization but also to the lack of a full examination of fibre retention kinetics and biodurability and the processes that affect these. Another gap relates to the use of cells and lung tissues from animals exposed *in vivo*, preferably by inhalation, for subsequent in-vitro mechanistic studies. As an example, a technique for isolation of type II alveolar epithelial cells for evaluation of specific mutations has been developed only recently.

A major debate concerning long-term in-vivo studies relates to the suitability of the methods of fibre administration, with the three main techniques being inhalation, intratracheal instillation and intracavity injection. Inhalation is suggested as being a natural method of exposure where the normal lung defences are operating. It permits an examination of biopersistence, pulmonary toxicity, fibrosis and carcinogenicity with relevance to both pulmonary carcinomas and mesotheliomas, although it should be remembered that the dimensions of respirable fibres are different between rats and humans. Intratracheal injection has the disadvantage of being an artificial exposure that can swamp normal lung defences. However, if this is accepted, the technique may be used to examine biopersistence, pulmonary toxicity and carcinogenicity. Intracavity injection is also an artificial method of administration that bypasses lung defences, and there is a limit to the maximal diameter of the fibres that can be studied. However, the technique may be of use to examine the mechanisms of development of mesotheliomas and the long-term persistence of fibres in these cavities compared to the lung. This has been little examined and is a major gap in our knowledge.

The major point of difference regarding the value of inhalation versus injection studies relates to their sensitivity and the applicability of results for predicting human hazard and risk. The point of major importance with respect to mechanisms of fibre carcinogenicity relates to the issue of dose, as mechanisms at high-dose levels may be different from those at low-dose levels. Although all techniques may be of use in examining particular aspects of the carcinogenic process from a mechanistic point of view, it needs to be considered that, after deposition in the respiratory tract, translocation of fibres from the alveolar region to the pleura represents a selection process in terms of fibre dose and fibre size. This aspect is circumvented with direct intracavity administration of fibres.

With regard to the development of fibre-related disease and tumour production in the presence of other non-fibrous dusts, chemical carcinogens, viruses and radiation, our knowledge of the effects of combinations such as these is almost one large gap. The few studies on these combinations that have been published have major weaknesses in study design, particularly their lack of proper controls. What is needed are a series of studies where fibres are administered with and without one of the materials under consideration in controlled multidose studies using a protocol where some level of fibre carcinogenicity is known to occur and including the collection of data on variations of fibre clearance or fibre durability that might occur with the different combinations.

Relevance of in-vitro assays

To what extent can physico-chemical properties be used to predict potential carcinogenicity of fibres?
At present, there is insufficient understanding of how the physical and chemical properties of fibres contribute to mechanisms of fibre-induced carcinogenesis to make reliable predictions of the carcinogenic potential of fibres based solely on these types of data. In this respect, it was considered that there were no combinations of physical and chemical data that could be used to identify a fibre as a carcinogen or a noncarcinogen. However, there are physical and chemical properties of fibres that have been associated with fibre toxicity *in vitro* and toxicity and/or carcinogenicity *in vivo*. In this respect, characterizing selected physical and chemical properties of fibres could be useful in the context of screening assays to make inferences on the relative potential of fibres to produce adverse effects *in vivo*. Given the current limitations of in-vitro fibre testing (see above), these inferences would need to be validated *in vivo*.

Fibre dose, dimensions and durability are currently accepted as important parameters; are there other important characteristics relevant to potential bioactivity?
In addition to dimension and durability, there may be other aspects of the physical and chemical properties of fibres that can provide information on potential fibre toxicity *in vivo*. These were considered to include the following: the presence of iron or other transition metals on fibres; the ability of a fibre to accumulate iron; the ability of fibres to generate free radicals; the ability of a fibre to interact with and alter biologically relevant molecules (e.g. DNA, lipids, proteins); and the ability of fibres to cause lysis of erythrocytes/liposomes. In addition to these endpoints, information on the ability of fibres to activate cells *in vitro* (e.g. to produce ROS and cytokines and/or alter the expression of proliferation-related genes of macrophages, epithelial cells, mesothelial cells and other relevant cells) may provide insights into the relative potential of fibres to elicit adverse effects *in vivo*. Since the precise relationships between these various aspects of fibre activity and potential chronic toxicity *in vivo* are incompletely understood, these data would be useful primarily in the context of screening assays and the need for chronic toxicity to be confirmed by appropriate in-vivo testing.

Are in-vitro genotoxicity assays relevant to fibre carcinogenesis?
It is generally agreed that genetic alterations play a critical role in the carcinogenic process. Thus, conceptually, well-designed and validated genotoxicity assays can provide information for assessing the potential carcinogenicity of materials.

In the context of genotoxicity testing of fibres, current in-vitro genotoxicity tests possess limitations common to all current in-vitro assays on fibres. In addition, questions exist regarding proper validation of genotoxicity tests for fibres in that there are no clear negative control materials, nor is there agreement against which in-vivo carcinogenicity data the tests should be compared. Recognizing these limitations, however, in-vitro genotoxicity assays, particularly those assessing cytogenetic effects, do provide some information on potential in-vivo genotoxicity of fibres and, in this regard, potential carcinogenicity. However, given the current limitations of in-vitro testing methods for fibres, this type of in-vitro information should be validated by appropriate in-vivo tests.

What is the relationship between acute in-vitro effects of fibres on growth factor and proto-oncogene expression and chronic persistent proliferation of target cell populations?
Fibres can persist within the lung, and studies have demonstrated fibre-specific differences in biopersistence. Conceptually, the biopersistence of fibres should relate to their ability to activate gene expression and induce cell proliferation over extended time periods. However, there are significant gaps in our knowledge of the relationships between growth factor/proto-oncogene expression in in-vitro or short-term in-vivo studies and cell proliferation. In addition, there are gaps in our understanding of how the effects of fibres on cell proliferation after acute exposure relate to cell proliferation after chronic exposure. Therefore, at present, the relevance of in-vitro and acute in-vivo changes in gene expression to potential chronic proliferative effects is uncertain.

Relevance of in-vivo assays

Can the genotoxic effects of fibres be assessed in vivo?
To date, there has been virtually no assessment of genotoxicity *in vivo*. In theory, it should be possible using such endpoints as unscheduled DNA synthesis, the occurrence of mutations *in vivo* and also the occurrence of cytogenetic abnormalities in target cell populations following exposure to fibres.

It will be technically difficult to undertake such studies with mesothelial cells due to the monolayer nature of the mesothelium and the associated difficulty in isolating sufficient numbers of these cells. For other target cells, validated separation techniques already exist (e.g. type II alveolar epithelial cells).

What are the links between inflammation, fibrosis and cancer induced by fibres?
Experimental studies with fibres showing significant numbers of lung tumours have always shown high levels of pulmonary fibrosis. This does not necessarily indicate a cause–effect relationship since both processes may be a response to high fibre doses. Intracavity injection studies using high doses result in both mesotheliomas and fibrosis. There is, however, debate concerning whether low but carcinogenic doses of fibres also result in fibrosis in this model. This may be due to the lack of sensitive methods for estimating fibrosis, particularly in the peritoneal cavity. The relationship between pleural fibrosis and mesothelioma has not been determined. It is not known whether malignant mesothelioma arises from the visceral pleura, the parietal pleura or both.

Fibre-induced chronic inflammation leads to fibrosis. There are no data on direct links between inflammation and carcinogenesis. However, one widely held theory is that, in areas of chronic inflammation, substances such as ROS and cytokines are produced that may be involved in tumour production.

Do lung burden and biopersistence of fibres in animals reflect lung burden and biopersistence of fibres in humans?
Sizes of 'respirable' fibres are different between rats and humans. This means that, for any dust cloud, penetration and deposition (and therefore lung burden accumulation) will be different. Once in the lung, differences in the clearance and translocation rates will also affect the retained lung burden. Few data exist on these matters, although human alveolar macrophage-mediated clearance of non-fibrous materials is slower than that for rats. The rates for clearance from the interstitium have not been compared. Fibre dissolution, which is a chemical process, may not differ substantially between species, although some work with cobalt oxide does show some interspecies differences.

Is total lung fibre burden an accurate assessment of fibre disposition or are there localized areas of fibre deposition and retention that correlate with the development of bronchogenic carcinoma and mesothelioma?
Localized accumulation of fibres in the form of fibre-containing lesions at the bifurcations of terminal and respiratory bronchioles certainly

Table 1. Summary of experimental endpoints after in-vitro and in-vivo exposure to fibres					
Experimental design	Oxidant-induced damage	Aneuploidy	Cell proliferation	Inflammation	Co-carcinogenicity
In vitro					
Rodent cell lines	++	++	+/–	++	+/–
Human cell lines	+	+/–	0	++	0
In vivo					
Rodents – short term	+	0	++	++	0
Rodents – long term	0	0	++	++	+/–
Humans	0	0	0	++	+/–

++, strong effect; +, weak effect; –, no effect; 0, no data; +/– contradictory data.
See authored papers elsewhere in this volume for a full discussion of these experimental data.

occur. Other possible 'hot spots' are areas of interstitial or peribronchiolar fibrosis and lymphatic stomata in the parietal pleura. However, no data exist as to whether dissolution or clearance from these sites of aggregation differs from the rest of the lung. In rats, pulmonary tumours often appear to develop in the proximity of 'hot spot' lesions. However, human bronchogenic carcinomas develop mainly in the larger proximal bronchial airways and not in the distal lung parenchyma.

Does the inhalation of fibres or mixed fibres and non-fibrous dusts impair clearance in rats? Is this mechanism relevant for humans?
Inhalation of fibres at high doses in experimental studies has been shown to produce an increased rate of accumulation in proportion to the mass of dust deposited. Information is lacking on the effects that mixtures containing non-fibrous dusts have on the accumulation or clearance of fibres. Fibres, themselves, have been shown to impair the clearance of some materials such as cobalt oxide at moderate doses.

It is not certain whether human asbestos exposures in the past were ever sufficiently high to produce impaired clearance.

Relevance of mechanistic data in evaluation of fibre carcinogenicity to humans
Cellular and molecular mechanisms of fibre carcinogenesis
The exact mechanisms leading to the development of cancer after exposure to asbestos fibres are poorly understood. Most lung cancers in humans exposed to asbestos occur in cigarette smokers; however, an excess of lung cancer also occurs in a small percentage of people exposed to asbestos fibres alone. It is not known whether the same mechanism is responsible for the development of these tumours in smokers and nonsmokers exposed to asbestos. While fibres can produce both lung cancer and mesothelioma, different patterns of molecular alterations have been identified in human lung cancers associated with asbestos exposure and cigarette smoking in comparison with diffuse malignant mesothelioma. Therefore, it is possible that different cellular and molecular mechanisms are involved in the development of these two tumour types. Specific molecular alterations unique to asbestos-induced tumours have not been identified in either humans or experimental animals. Therefore, it is difficult to assess whether similar molecular mechanisms are responsible for the development of lung cancer and mesothelioma in humans and rodents exposed to asbestos fibres. Similarly, it is not known whether different types of carcinogenic fibres activate common or different mechanistic pathways.

As summarized in the paper by A.B. Kane in this volume, five mechanistic hypotheses for fibre carcinogenicity have been proposed:

- Fibres generate free radicals that damage DNA.
- Fibres interfere physically with mitosis.
- Fibres stimulate proliferation of target cells.

- Fibres provoke a chronic inflammatory reaction leading to prolonged release of ROS, cytokines and growth factors.
- Fibres act as co-carcinogens or carriers of chemical carcinogens to the target tissue.

This Consensus Report has summarized the strengths, weaknesses and gaps in the published data that are relevant for these hypotheses; these observations are summarized in Table 1. It should be noted that some of the experimental endpoints listed in this table have also been noted after exposure to non-fibrous particles. These experimental observations reveal associations between exposure to asbestos fibres and specific endpoints in in-vitro or in-vivo models. Some of these associations have also been observed in people exposed to asbestos fibres. However, few experiments have been conducted to assess critically the causal relationship between these changes and the development of lung cancer or mesothelioma.

The inflammatory endpoints measured in various in-vitro and in-vivo assays are especially pronounced at high-dose exposures. It is not known whether tumours produced by high-dose exposures develop via similar or different mechanisms in comparison with low-dose exposures. Most in-vitro studies have been conducted at relatively high fibre : cell ratios; the relevance of these data for chronic exposures at lower doses *in vivo* is also questionable.

Overall, the available evidence in favour of or against any of these mechanisms leading to the development of lung cancer and mesothelioma in either animals or humans is evaluated as weak.

Recommended experimental studies

Future evaluations of fibre carcinogenicity where human epidemiological data or chronic inhalation assays are limited or not available will depend in part on mechanistic information based on relevant experimental models. The Workshop concluded that the following experimental studies would provide additional data for future evaluation of fibres:

- A multidose, chronic inhalation study in rats and hamsters using a well-characterized amphibole sample; this study should include relevant short-term endpoints or biomarkers that could be evaluated in future mechanistic studies.

- New in-vitro models including development of systems to evaluate the effects of fibre dissolution and in-vivo/ex-vivo assay systems, especially for evaluation of the potential genotoxic and clastogenic effects of fibres.

References

Bignon, J., Saracci, R. & Touray, J.-C., eds (1994) *Biopersistence of Respirable Synthetic Fibres and Minerals.* (*Environ. Health Perspectives*, Vol. 102, Suppl. 5)

WHO/EURO Technical Committee for Monitoring and Evaluating Airborne MMMF (1985) *Reference Methods for Measuring Airborne Man-made Mineral Fibres (MMMF)*, Copenhagen, World Health Organization Regional Office for Europe

Mechanisms of mineral fibre carcinogenesis

A.B. Kane

Introduction

Human diseases resulting from inhalation of fibres

The lower respiratory tract is the major target of inhaled particles and fibres with diameters less than or equal to 3 µm (McClellan et al., 1992). The adverse health effects of inhaled fibres were first recognized in asbestos workers and people living in the vicinity of asbestos mines. Asbestos-related diseases were described in isolated case reports, beginning with asbestosis in 1924, and were followed by cohort mortality studies of lung cancer published by R. Doll in 1955 and malignant mesothelioma published by J.C. Wagner and co-workers in 1960 (for an historical review see Gordon, 1992).

The human diseases associated with exposure to asbestos fibres are summarized in Table 1. The conducting airways of the respiratory tract, the alveolar sacs, where gas exchange takes place, and the pleural linings surrounding the lungs are sites of these asbestos-related diseases. The spectrum of asbestos-related diseases ranges from nonspecific effects caused by any inhaled irritant, to fibrosis or scarring by collagen deposition, to cancer. The clinical and pathological features of these diseases will be summarized briefly below. The neoplastic diseases specifically associated with exposure to asbestos fibres will be emphasized. Pathological reactions associated with exposure to other natural and man-made fibres will be included, where this information is available.

Airway diseases. The major airways of the lower respiratory tract are the site of chronic bronchitis and chronic limitation of airflow, which are caused by persistent inflammation and excess mucus secretion. These are nonspecific reactions to various pollutants, including noxious gases, particulates and cigarette smoke. In the smaller, more distal airways, particulates accumulate in the walls of the terminal respiratory bronchioles. This is also a nonspecific reaction that occurs in cigarette smokers. These airway diseases cause increased morbidity, but they are not usually lethal (Becklake, 1994).

Asbestosis. Diffuse, bilateral interstitial fibrosis is the thickening of the walls of the alveolar sacs by increased deposition of connective tissue. Similar to the airway diseases associated with asbestos exposure, this pattern of fibrotic scarring of the lungs is nonspecific and occurs in response to a variety of insults to the alveoli. Asbestosis is characterized by fibrosis in the subpleural regions of the lower lobes of the lungs. Histopathological examination of lung tissue reveals the presence of asbestos bodies – fibres coated with haemosiderin, protein and mucopolysaccharides. Asbestosis is a progressive disease with clinical signs and symptoms developing after 10 or more years of

Table 1. Human diseases associated with asbestos exposure

Malignant diseases[a]	Non-malignant diseases[b]
Lungs	
Bronchogenic carcinoma	Asbestosis
	Asbestos-related small airway disease
	Major airway diseases: chronic bronchitis, chronic airflow limitation
Pleura	
Diffuse malignant mesothlioma	Effusion
	Pleural plaques
	Diffuse visceral pleural fibrosis
	Rounded atelectasis

[a]Churg & Green, 1995.
[b]Becklake, 1994.

repeated exposure to asbestos fibres. The disease was more prevalent in the past when exposures were several orders of magnitude greater than current exposures. Those affected gradually develop shortness of breath and impaired gas exchange leading to the inability to work, respiratory failure, heart failure (cor pulmonale) and premature death (Craighead et al., 1982). Patients with asbestosis are at high risk of developing bronchogenic carcinoma, especially if they smoke cigarettes (Churg & Green, 1995); however, a causal relationship between fibrosis and bronchogenic carcinoma has not been established.

Bronchogenic carcinoma. Lung cancer arises from the epithelial lining of the large and small airways of the lungs. In most populations, cigarette smoking is the most common cause of lung cancer. Both nonsmokers and smokers exposed to asbestos fibres may develop bronchogenic carcinoma, although the risk is greatly increased in smokers. Similar to asbestosis, there is a latent period of at least 10–20 years between exposure to asbestos and the clinical manifestations of bronchogenic carcinoma. These cancers are associated with persistent cough, recurrent pneumonia, bleeding, weight loss or symptoms associated with metastatic spread. Bronchogenic carcinomas frequently metastasize early; survival beyond two to five years is rare (Cagle, 1995). Environmental exposure to erionite fibres has also been associated with the development of bronchogenic carcinoma (IARC, 1987a).

It is hypothesized that a single epithelial cell precursor gives rise to the four major histological types of bronchogenic carcinoma: small-cell carcinoma, large-cell carcinoma, squamous-cell carcinoma and adenocarcinoma (Cagle, 1995). An important subtype of adenocarcinoma is bronchiolo-alveolar carcinoma because it has unique clinical and pathological characteristics; these tumours arise in the periphery of the lungs and they may be multifocal. The incidence of bronchiolo-alveolar carcinoma appears to be increasing. This type of lung cancer frequently occurs adjacent to a peripheral, fibrotic scar; it has been described as 'scar carcinoma' (Barsky et al., 1994). However, histopathological studies suggest that this type of cancer does not arise in a pre-existing scar; rather it stimulates a host fibrotic or desmoplastic reaction leading to retraction of the pleura overlying the growing neoplasm (Barsky et al., 1986). This is an important observation because fibrous scarring is considered to be a predisposing factor for lung cancer in workers exposed to asbestos. The relationship between fibrosis and carcinogenesis will be discussed below (see *Mechanisms of fibre carcinogenesis*). Previously, adenocarcinomas were thought to be more common in asbestos workers than other histological types of lung cancer; however, most pathologists now agree that all histological types of lung cancer occur in association with asbestos exposure (Mollo et al., 1990; Cagle, 1995).

Specific histological lesions develop in the respiratory epithelium prior to the appearance of a malignant carcinoma. These lesions are described as metaplasia, hyperplasia, atypical hyperplasia, dysplasia and carcinoma *in situ*. Specific genetic alterations may be identified in these pre-neoplastic lesions (Cagle, 1995); the specificity of these genetic changes in relationship to asbestos and cigarette smoking will be discussed below (see *Mechanisms of fibre carcinogenesis*).

Non-malignant pleural diseases. A monolayer of flat mesothelial cells covers both the surface of the lungs (visceral pleura) and the inner chest wall and the diaphragm (parietal pleura). Both of these layers of the pleura are affected by exposure to asbestos fibres, although the mechanism responsible for these reactions is unknown. The most common reaction to asbestos fibres is parietal pleural plaque. This is a fibrotic scar that develops on the lateral chest walls or on the superior surface of the diaphragm. These scars may become calcified and therefore visible on a chest X-ray, usually after 20–30 years of occupational asbestos exposure. People living in environments where asbestos fibres are present in the soil or near asbestos mines and industries also develop parietal pleural plaques (Bignon, 1989). Some physicians consider parietal pleural plaques to be biomarkers of asbestos exposure and a warning signal that these workers are at a higher risk of developing subsequent malignant complications (Hillerdall, 1994).

Less commonly, asbestos workers develop recurrent episodes of fluid accumulation in the space between the visceral and parietal layers of the pleura; this is called pleural effusion or benign asbestos pleurisy. The fluid may compress the lungs and require drainage, and in some patients, may lead to diffuse fibrosis or scarring of the visceral pleura. Rounded atelectasis is a focal area of compression and collapse of the lung parenchyma due to an adjacent fibrous scar of the visceral pleura. Although none of these lesions is a precursor of a malignant neoplasm, these fibrotic pleural lesions may impair normal lung function (Schwartz, 1991).

Malignant pleural disease. Diffuse malignant mesothelioma is a lethal neoplasm arising from the pleura, the peritoneum or, rarely, the pericardium or tunica vaginalis of the testis (Churg & Green, 1995). The routes of translocation of asbestos fibres to these sites are unknown, although Oberdörster (1994) discusses experimental evidence for lymphatic dissemination of amphibole fibres. Approximately 80–85% of mesothelioma cases are associated with a history of occupational exposure to asbestos, especially to amphibole asbestos. Malignant mesothelioma has the longest latent period of all of the asbestos-related diseases – up to 50–60 years after the first exposure. No other known cofactors contribute to this malignant neoplasm; in contrast to bronchogenic carcinoma, there is no increased incidence in cigarette smokers (Churg & Green, 1995). Pleural plaques and malignant mesothelioma have also been found in people exposed to asbestiform fibres, especially erionite, or to tremolite in the environment (reviewed by Bignon, 1989).

Malignant mesothelioma arising in the pleural lining usually presents clinically as a diffuse mass encasing the lung in the presence of a bloody pleural effusion. There may be local invasion of the lung or chest wall, followed by metastases via the lymphatic system or bloodstream. Most patients have an extensive tumour mass upon diagnosis and suffer from difficulty in breathing, chest pain, weight loss, cough and fever (Musk & Christmas, 1992). Surgical resection of the tumour is difficult and malignant mesothelioma is resistant to chemotherapy and radiation. Few patients survive more than one to two years (Churg & Green, 1995).

Malignant mesotheliomas have a wide range of histological patterns; they can resemble epithelial carcinomas or fibroblastic sarcomas or a mixture of both. Pathologists must be careful to distinguish between a primary epithelial type of malignant mesothelioma and a metastatic adenocarcinoma. A variety of specialized tumour markers are used to aid in this difficult differential diagnosis (Henderson *et al.*, 1992). Specific molecular markers (see *Molecular alterations in asbestos-related neoplasms*) may eventually be useful in the diagnosis of primary malignant mesothelioma. In most patients, this neoplasm has reached an advanced stage at the time of clinical diagnosis; therefore, no specific histopathological precursor lesions have yet been identified. Examination of the pleural surfaces by thoracoscopy is a newer technique that may lead to identification of specific precursor lesions and earlier diagnosis of this neoplasm in workers exposed to asbestos (Boutin & Rey, 1993). These recent investigations provide evidence that mesothelioma may arise from the parietal pleura.

Other malignant diseases. Asbestos fibres have been found as contaminants of water and beverages; this observation leads to the possibility that this contamination may contribute to carcinomas of the gastrointestinal tract. Multiple epidemiological studies have not provided strong evidence for this association (reviewed in Bignon, 1989). In general, there is no accepted relationship between asbestos exposure and neoplasms other than bronchogenic carcinoma and malignant mesothelioma (Churg & Green, 1995).

In-vitro and in-vivo models
A variety of in-vitro and in-vivo experimental models have been developed to study the pathogenesis of asbestos-related diseases (see Tables 2 and 3). The accompanying reviews in this volume summarize recent experimental data obtained from these models; the limitations of these in-vitro and in-vivo model systems are discussed (see Consensus Report). Recent reviews of the in-vitro and in-vivo effects of man-made fibres have been published (Wheeler, 1990; Ellouk & Jaurand, 1994).

A central issue in extrapolating results from in-vitro to in-vivo models is the determination of the dose delivered to the target tissue. Investigators are now beginning to express dose in terms of fibre number rather than mass; a further improvement would also be the determination of the number of

Table 2. In-vitro models developed to study the pathogenesis of asbestos-related dieases

Model	Endpoints
Cell-free models	DNA damage
	Lipid peroxidation
Target cell populations[a]	Cell toxicity and apoptosis
	Release of inflammatory mediators
	Genotoxicity
	Transformation
	Cell proliferation
	Metaplasia
	Gene expression
	Intercellular communication

[a]Macrophages; lung epithelial cells; lung fibroblasts; mesothelial cells; organ cultures.

fibres delivered to target cells *in vitro* (Oberdörster, 1994) or *in vivo* (MacDonald & Kane, 1986). However, extrapolation of the dose delivered to target cells in these experimental models to the dose delivered to humans exposed to natural and man-made fibres is a major challenge (McClellan & Hesterberg, 1994).

Future epidemiological studies of workers exposed to recently developed fibres will be limited by the following factors. First, there will be no

Table 3. In-vivo models developed to study the pathogenesis of asbestos-related diseases

Model	Endpoints
Short-term animal models	Fibre deposition
	Inflammation
	Cell proliferation
	Pre-neoplastic changes
Long-term animal models	Lung and pleural fibre burdens
	Fibrosis
	Tumours
	Molecular and cytogenetic alterations

adequate epidemiological studies with sufficient latency. Second, the levels of exposure to these man-made fibres will almost certainly be lower than past exposures to asbestos and the worker populations are likely to be smaller. Therefore, it will be necessary to rely on experimental data based on in-vitro and in-vivo models in future evaluations of natural and man-made fibres.

Classification and characterization of fibres that are known or potential carcinogens for humans
Classification of natural and man-made fibres
Fibrous minerals are divided into two major categories: asbestos and asbestiform (see Table 4). Asbestos is a term used to describe hydrated fibrous silicates, although the crystalline structure of serpentine asbestos is different from that of the amphiboles (IARC, 1977). Three types of asbestos fibres are used commercially: chrysotile, amosite and crocidolite. The other amphiboles may be present in chrysotile, vermiculite or talc mines. Three types of asbestiform fibres are used commercially: attapulgite, sepiolite and wollastonite (for more details see: Bignon, 1989; IARC, 1987a). People are exposed to the asbestos and asbestiform fibres that are used commercially; exposure to natural deposits of these fibres at the soil surface also occurs in the general population, even in rural areas. The magnitude of non-occupational exposures to natural fibres is usually very low and difficult to quantify (Bignon, 1989).

Many man-made fibres have been produced that resemble natural asbestos and asbestiform fibres in their geometry and flexibility, although man-made fibres do not generally split longitudinally. Man-made fibres are classified as vitreous or crystalline (see Table 5) and vary considerably in their chemical composition, strength, temperature resistance and durability. Fibres have also been derived from organic sources and their potential pathogenicity should also be assessed (Davis, 1992).

IARC evaluations of fibres as carcinogens
The experimental evidence for asbestos and asbestiform fibres as human carcinogens was evaluated by IARC working groups in 1977 and 1987 (IARC, 1977, 1987a,b, 1988; see Table 6). All forms of asbestos, talc containing asbestiform fibres and erionite fibres were evaluated as known human carcinogens (Group 1). Other naturally occurring fibrous

silicates were placed in Group 3 (not classifiable). Since these IARC evaluations, occupational exposure to anthophyllite has been associated with the development of diffuse malignant mesothelioma (Meurman et al., 1994). The evidence for the carcinogenicity of five types of man-made vitreous fibres (MMVF) was evaluated by an IARC Working Group in 1988. Glasswool, rockwool, slagwool and ceramic fibres were placed in Group 2B (possibly carcinogenic to humans); glass filaments were placed in Group 3 (not classifiable).

Recognition of the diseases associated with past exposure to asbestos fibres raises concern about the potential carcinogenicity of other natural and man-made fibres. As of 1989, at least 70 different types of man-made fibres had been developed (Bignon, 1989). Since the previous IARC classifications of natural (IARC, 1977, 1987b) and man-made fibres (IARC, 1988), new experimental data have been obtained about the potential mechanisms of asbestos carcinogenesis. Concurrently, the procedure of IARC working groups has been modified to consider more explicitly certain mechanistic information in the assessments of the carcinogenic risk of agents for humans. Examples of mechanistic data that can be used in the evaluations are genotoxicity, effects on gene expression, cellular and tissue interactions and evidence of time and dose relationships in multistage models of carcinogenesis (Vainio et al., 1992). Recent experiments that have provided new mechanistic data relevant to fibre carcinogenesis are summarized in the accompanying authored papers in this volume. The amount of experimental data available has expanded considerably since 1989, and these data have also provided important information about the physico-chemical properties of fibres that may contribute to their biological activity. Identification of these critical physico-chemical parameters may provide clues about the mechanisms responsible for fibre carcinogenicity.

Properties of fibres relevant to biological activity
Dimensions. Fibre length was the first physico-chemical property to be associated with the carcinogenic potential of various natural and man-made fibres. This observation was reported by Pott and Friedrichs (1972) and by Stanton and Wrench (1972) using a rat model of malignant mesothelioma produced by direct intraperitoneal or intrapleural injection of fibres. Fibres longer than 8 μm in length

Table 4. Classification of natural fibres[a]

Asbestos	Asbestiform
Serpentine	**Fibrous clays**
Chrysotile	Palygorskite (attapulgite)
	Sepiolite
Amphiboles	**Other fibrous silicates**
Actinolite	Wollastonite
Amosite	Nemalite (fibrous brucite)
Anthophyllite	Talc
Crocidolite	Zeolites:
Tremolite	• mordenite
	• erionite

[a]Modified from Bignon, 1989.

and thinner than 0.25 μm in diameter were found to be more potent in inducing mesothelioma than shorter fibres of the same chemical composition (Stanton et al., 1981). Although this conclusion has been debated (for example, by Dunnigan, 1984; Goodglick & Kane, 1990), most investigators agree that long fibres are more carcinogenic than short fibres. The mechanism responsible for the increased

Table 5. Classification of man-made fibres

Vitreous (MMVF)[a]
Glasswool
Glass filaments
Rockwool
Slagwool
Ceramic fibres

Crystalline[b]
Alumina
Graphite
Potassium titanate
Silicon carbide
Sodium aluminum carbonate
Synthetic zeolites

Organic
para-Aramid
Cellulose

[a]IARC, 1988.
[b]Modified from Leineweber, 1980.

Table 6. IARC evaluations of the strength of the evidence from human and experimental animal data for the carcinogenicity of fibres

Agent	Evidence from Humans	Evidence from Animals	Overall evaluation[a]
Asbestos[b,c]	Sufficient	Sufficient	1
Silicates[d]			
Wollastonite	Inadequate	Limited	3
Attapulgite	Inadequate	Limited	3
Sepiolite	No data	Inadequate	3
Talc			
• not containing asbestiform fibres	Inadequate	Inadequate	3
• containing asbestiform fibres	Sufficient	Inadequate	1
Erionite	Sufficient	Sufficient	1
Man-made vitreous fibres (MMVF)[e]			
Glasswool	Inadequate	Sufficient	2B
Glass filaments	Inadequate	Inadequate	3
Rockwool	Limited	Limited	2B
Slagwool	Limited	Inadequate	2B
Ceramic fibres	No data	Sufficient	2B

[a]1, the agent is carcinogenic to humans; 2B, the agent is possibly carcinogenic to humans; 3, the agent is not classifiable as to its carcinogenicity to humans.
[b]Asbestos: all forms – actinolite, amosite, anthophyllite, chrysotile, crocidolite, tremolite.
[c]IARC, 1977, 1987b.
[d]IARC, 1987a.

potency of long fibres is unknown; Jaurand (1989) summarizes the following hypotheses:

- short fibres are phagocytized more readily by macrophages and thereby cleared more efficiently from the lungs;
- long fibres are incompletely phagocytized and trigger the release of more reactive oxygen/nitrogen species (ROS) than do short fibres;
- long fibres interfere with mitosis and chromosome segregation in dividing target cell populations.

These mechanisms will be reviewed in detail subsequently. Fibre diameter is also postulated to be an important parameter in carcinogenicity; for example, the transformation of Syrian hamster embryo cells (SHE) *in vitro* is sensitive to fibre diameter as well as fibre length (Hesterberg & Barrett, 1987).

Rodents were subsequently exposed to sized preparations of asbestos fibres by inhalation, and these experiments also confirmed the importance of fibre length in inducing pulmonary fibrosis and tumours (reviewed by Davis, 1994). The exact length of fibres responsible for these pulmonary reactions is uncertain. However, longer fibres are considered to be important in the pathogenesis of bronchogenic carcinoma based on the observation that fibres longer than 10 µm preferentially deposit at large airway bifurcations in casts of human airways (Sussman *et al.*, 1991). In animal inhalation studies and in humans exposed to fibres by inhalation, fibre dimension is a critical parameter in respirability. The dimensions of respirable fibres varies between humans and different species of experimental animals, and this variation must be considered in the design and interpretation of animal inhalation studies (McClellan & Hesterberg, 1994). Fibre dimension also contributes to biopersistence in the lung, as discussed below.

Chemical composition. Magnesium and iron, respectively, are important cations of the crystalline framework of serpentine and amphibole types of asbestos (IARC, 1977; Leineweber, 1980). Surface magnesium ions are removed from chrysotile asbestos by acid-leaching *in vitro* or more slowly by progressive leaching *in vivo*; this alteration in cation composition has been observed to decrease the cytotoxicity and carcinogenicity of chrysotile asbestos (Monchaux et al., 1981). The extent of chemical leaching of chrysotile in human lungs is controversial (Churg, 1994). Ferric and ferrous cations are major components of the crystalline lattice of amphibole asbestos fibres; iron may also be present as surface impurities on serpentine asbestos, amphiboles or some man-made fibres. The availability of iron at the surface of fibres is a critical parameter in catalysing the generation of ROS (reviewed by Fubini, 1993). Asbestos and other mineral fibres such as erionite may release or acquire iron from the surrounding medium, depending on the presence of chelators or reducing agents (Hardy & Aust, 1995; Eborn & Aust, 1996). The potential role of the iron-catalysed generation of ROS in asbestos-related cancers will be discussed below (see *Mechanisms of fibre carcinogenesis*).

Surface reactivity. Surface area is an important parameter that contributes to the surface reactivity of fibres. The unique structure of fibres such as erionite greatly increases their surface area; internal 'cages' in erionite and synthetic zeolites function as cation exchangers (Leineweber, 1980). Chemical composition, surface charge and surface area also influence the ability of fibres to adsorb exogenous or endogenous molecules. Surface adsorption of exogenous compounds such as polycyclic aromatic hydrocarbons (PAH) may be important in the pathogenesis of bronchogenic carcinoma, as discussed below. Adsorption of endogenous macromolecules such as lung surfactant or immunoglobulins may alter the biological reactivity of fibres (Scheule & Holian, 1989), and lipids, proteins and DNA may also adhere to the surface of fibres (Ellouk & Jaurand, 1994).

Durability and biopersistence. Durability is an important characteristic of fibres that are used commercially. Durability is usually assessed by dissolution *in vitro*. Biopersistence describes the overall retention of fibres in the respiratory tract, and includes mechanical clearance, dissolution and leaching. This is a complex process dependent on the local extracellular and intracellular environment. Fibres may split longitudinally or transversely after deposition in the lungs. Splitting of chrysotile fibres may initially increase the number of fibres in the lungs (Davis, 1994); however, eventually, the reduced length of the fibres increases their rate of clearance. Clearance also depends on lung fibre burden, as described recently for refractory ceramic fibres (Yu et al., *1994)*. The chemical composition and surface architecture of MMVFs were shown to be altered dramatically after chronic inhalation in rats (Hesterberg et al., 1994). In contrast, in this same model, refractory ceramic fibres were altered less in rats and hamsters (Mast et al., 1994). Other fibre types such as wollastonite have been shown to persist for very short periods after short-term inhalation by rats (Warheit et al., 1994). Amphibole fibres persist longer than chrysotile in human lungs (reviewed by Case, 1994; Churg, 1994). Few studies have investigated the persistence of man-made fibres in people. According to Sébastien (1994), glass fibres may be less persistent than chrysotile fibres. There are insufficient data for mineral wool or ceramic fibres, but one study reports prolonged persistence of silicon carbide fibres (Sébastien, 1994).

The physical and chemical modification of fibres in the lungs is hypothesized to reduce their biopersistence and biological reactivity (reviewed by Davis, 1994; De Vuyst, 1994). However, certain caveats have been raised about this generalization. First, the following 'hot spots' of prolonged fibre persistence have been proposed: (i) sites of lymphatic drainage in the parietal pleura (De Vuyst, 1994); (ii) airway bifurcations (Lippman, 1994); and (iii) within areas of fibrosis (Davis, 1994). Second, different fibre types or modified fibres may alter the mobility of macrophages and the translocation of fibres towards the pleura or lymph nodes (Davis, 1994). Finally, Barrett (1994) warns that no relationship has been established between biopersistence of fibres in the lung and the induction of genetic and epigenetic changes that may lead to cancer.

It is important to consider these caveats in the interpretation of internal fibre dose (which is measured as the number of fibres that persist in the lungs – the lung fibre burden – at the time of diagnosis of disease or at autopsy of exposed people). Although this information is important, especially in corroborating exposure to specific types of fibres in the workplace, the measurement and interpretation of human lung fibre burdens is a complex issue (reviewed by Becklake & Case, 1994; Sébastien, 1994).

Unanswered questions. Given our current knowledge about the properties of fibres relevant for biological activity, the following questions need to be addressed (see Consensus Report):

- To what extent can physico-chemical properties be used to predict the potential carcinogenicity of fibres? Fibre dimensions and durability are currently accepted as important parameters; are there other important characteristics?
- Does the biopersistence of fibres in animals reflect the biopersistence of fibres in humans?
- Is total lung fibre burden an accurate assessment of fibre disposition, or are there localized areas of fibre deposition and retention that correlate with the development of bronchogenic carcinoma and malignant mesothelioma?

Molecular alterations in asbestos-related neoplasms

Malignant mesothelioma

Malignant neoplasms are characterized by autonomous proliferation and multiple alterations in oncogenes and tumour suppressor genes (reviewed by Harris, 1992). Exposure to asbestos fibres may contribute directly or indirectly to these molecular alterations, as hypothesized by Walker et al. (1992). Alterations in growth regulatory genes, oncogenes and tumour suppressor genes have been investigated in human and rodent mesothelial cell lines, although few investigations have been carried out directly on primary tumours. Multiple alterations in oncogenes and tumour suppressor genes have also been described in various histological types of human bronchogenic carcinomas (reviewed by Viallet & Minna, 1990); however, few investigations have been carried out in patients who developed bronchogenic carcinoma following asbestos exposure alone or in combination with cigarette smoking. Three categories of molecular alterations have been investigated in human malignant mesothelioma cell lines: (i) alterations in oncogenes and growth factors; (ii) alterations in growth regulatory genes; and (iii) alterations in tumour suppressor genes. These experimental data will be summarized briefly.

Oncogenes and growth factors. Several investigators have examined human malignant mesothelioma cell lines for the activation of common oncogenes by point mutation or amplification: no changes have been found yet in the *ras* gene family. A novel oncogene has been discovered in a human mesothelioma cell line using an NIH 3T3 transfection assay; however, its sequence has not yet been determined (Walker et al., 1992). Some oncogenes encode for growth factors or their receptors; activation of these oncogenes up-regulates cell proliferation in neoplastic cells. The c-*sis* proto-oncogene that encodes for the B chain of platelet-derived growth factor (PDGF-B) is expressed in malignant, but not normal, human mesothelial cell lines. However, over-expression of PDGF-A chain appears to be more important for the autonomous proliferation of human mesothelial cells as determined by transfection studies using an immortalized mesothelial cell line (Van der Meeren et al., 1993). A second autocrine growth factor loop involving transforming growth factor-α (TGF-α) has also been reported in human mesothelioma cell lines (Morocz et al., 1994). No evidence for the amplification of the genes encoding for the epidermal growth factor receptor or the *met* growth factor receptor was found in human malignant mesotheliomas (Tiainen et al., 1992).

Growth regulatory genes. Cell proliferation in mammalian cells is controlled by the expression of cyclins and cyclin-dependent kinases, which are turned on and off in a precise temporal sequence during the cell cycle. Over-expression of a cyclin would result in continuous stimulation of the cell cycle and increased cell proliferation. Cyclin D1 is an important regulator of the G1 phase of the cell cycle; over-expression of the cyclin D1 protein has been reported in a series of human cancer cell lines, including two malignant mesothelioma cell lines (Schauer et al., 1994).

Cyclins, as well as cyclin-dependent kinases, function to drive cell cycle progression. Cyclin-dependent kinase inhibitors function as brakes, opposing the action of kinases. An important growth inhibitor is cyclin-dependent kinase 4 inhibitor (p16). This growth inhibitor has been found to be mutated or deleted in the majority of human carcinoma cell lines, including malignant mesotheliomas (Gerwin, 1994). These recent investigations provide evidence for the autocrine activation of at least two growth factor pathways (PDGF and TGF-α), in addition to the inactivation of p16, an important regulator of the G1 phase of the cell cycle. Together, these alterations could contribute to the autonomous or unregulated proliferation of malignant mesothelioma cells.

Tumour suppressor genes. Three tumour suppressor genes have been investigated in human malignant mesotheliomas: *Rb* (the retinoblastoma gene), *WT1* (the Wilm's tumour gene) and *p53* (the most commonly altered tumour suppressor gene, mutated in approximately 50% of human neoplasms). No alterations have been found in *Rb* in human mesothelioma cell lines; this finding is not surprising because over-expression of cyclin D1 may substitute for deletion of *Rb*. *WT1* is expressed in both normal and malignant human mesothelial cell lines; so far, no deletions of this tumour suppressor gene have been found in human malignant mesotheliomas, although more subtle alterations in this gene may be responsible for altered growth factor responses in these neoplasms (Gerwin, 1994). Two groups of investigators have examined human malignant mesothelioma cell lines for alterations in the tumour suppressor gene *p53*. Overall, only one-third of the cell lines investigated showed any alterations. In contrast to most human bronchogenic carcinomas, point mutations in the highly conserved exons 5–8 of this gene were rare (Cote *et al.*, 1991; Metcalf *et al.*, 1992).

The infrequent association of point mutations in *p53* with human malignant mesothelioma conflicts with the results of immunohistochemical assays to detect expression of p53 protein in human mesotheliomas *in situ*. Between 25 and 45% of these human neoplasms show increased expression of p53 (Mayall *et al.*, 1992; Ramael *et al.*, 1992). Increased p53 expression is usually associated with point mutations (in *p53*) that prolong the half-life of this nuclear phosphoprotein. In the absence of point mutations, other mechanisms must be responsible for the elevated level of p53 expression in human malignant mesotheliomas.

Bronchogenic carcinoma
The genetic lesions associated with the activation of oncogenes or inactivation of tumour suppressor genes may be indicative of exposure to specific chemical carcinogens (Hollstein *et al.*, 1991). For example, guanine → thymine transversions are induced by carcinogens in cigarette smoke; this mutation occurs frequently at codon 12 of the K-*ras* proto-oncogene in adenocarcinomas of the lung arising in smokers and ex-smokers (Westra *et al.*, 1993). The same mutation at this locus has also been shown to occur in lung adenocarcinomas of workers exposed to both asbestos and cigarette smoke (Husgafvel-Pursiainen *et al.*, 1993). Specific carcinogens in cigarette smoke most probably initiate this mutation in K-*ras* at an early stage in the development of lung adenocarcinomas (Westra *et al.*, 1993).

Accumulation of p53 has also been detected in bronchogenic carcinomas arising in patients exposed to asbestos and cigarette smoke. Also, as demonstrated by immunohistochemistry, p53 is over-expressed in lung carcinomas associated with cigarette smoking alone or in combination with asbestos exposure (Nuorva *et al.*, 1994). So far, the pattern of *p53* mutations in bronchogenic carcinomas associated with asbestos exposure alone or in combination with cigarette smoke has not been analysed. In cigarette smokers, over-expression of p53 resulting from point mutations in exons 5–8 is found in pre-neoplastic and dysplastic lesions adjacent to invasive neoplasms; thus, mutations in this tumour suppressor gene are also considered to be an early event in the development of bronchogenic carcinoma (Cagle, 1995).

Additional comparative studies between cell lines and archival tissue samples of experimental and human lung tumours associated with asbestos exposure are required to evaluate the role of mineral fibres in the induction of molecular alterations during the development of bronchogenic carcinoma and malignant mesothelioma.

Mechanisms of fibre carcinogenesis
Some clues about the mechanisms responsible for asbestos-induced carcinogenesis have been provided by (i) the physico-chemical properties of fibres known to be carcinogenic for humans and (ii) recent cellular and molecular investigations of human malignant mesothelioma and bronchogenic carcinoma. Although comparative studies of the

effects of asbestos and other natural and man-made fibres in in-vitro and in-vivo models are incomplete, the following five hypotheses are proposed.

Hypothesis 1: Fibres generate free radicals that damage DNA

Natural and man-made fibres have been shown to catalyse the generation ROS in cell-free or in-vitro model systems (reviewed by Kamp et al., 1992), and these ROS may indirectly cause genetic damage. The following mechanisms are proposed for generation of oxidants. First, fibres may directly participate in the iron-catalysed generation of ROS in the presence of molecular oxygen, superoxide anion or hydrogen peroxide. The hydroxyl radical (OH•) is the end-product of these reactions and it has been shown to modify DNA bases in a cell-free system (Leanderson et al., 1988) and an in-vitro cell model system (Takeuchi & Morimoto, 1994). Asbestos fibres have also been shown to catalyse iron-dependent oxidation of membrane lipids (Gulumian & Kilroe-Smith, 1987; Goodglick et al., 1989) producing reactive alkoxyl radicals (Kamp et al., 1992). Second, independently of iron or oxygen, asbestos fibres can also catalyse oxidation of PAH to a free radical (Graceffa & Weitzman, 1987). Third, fibres may activate phagocytes to release ROS. It is hypothesized that 'frustrated phagocytosis' of long fibres is a potent stimulus for release of ROS (Mossman & Marsh, 1989). After release of hydrogen peroxide, superoxide anion or nitric oxide from activated phagocytes, additional radicals may be generated including OH•, peroxynitrite and nitronium ions (Ohshima & Bartsch, 1994). It is hypothesized that chronic generation of these various types of free radicals at sites of fibre deposition or during a persistent chronic inflammatory reaction mediates DNA damage that leads to critical mutations in oncogenes, growth regulatory genes and tumour suppressor genes. The evidence for and against this hypothesis that fibres generate free radicals that damage DNA will be summarized briefly below.

Assays for DNA damage. Most of the experimental studies searching for evidence of DNA damage induced by fibres have been conducted with various target cell populations *in vitro*. As reviewed by Jaurand (1989, 1991), the evidence for DNA damage is conflicting depending on the following variables: (i) fibre source and preparation, (ii) cell type, (iii) species and (iv) assay conditions. In general, asbestos fibres are considered to be inactive or weak inducers of point mutations or sister chromatid exchange. Direct measurement of DNA strand breaks has also been unsuccessful. However, asbestos fibres have been shown to induce multilocus deletions in a human–hamster hybrid cell line (Hei et al., 1992) and a variety of natural and man-made fibres have been shown to be clastogenic for rodent cells (Dong et al., 1994), but less damaging to human cells (Kinnula et al., 1995). Mesothelial cells appear to be more sensitive to DNA damage induced by asbestos fibres than do epithelial cells or fibroblasts; this increased sensitivity may be related to different levels of antioxidant defence mechanisms (Kinnula et al., 1992).

Cellular responses to DNA damage. Although direct measurements of DNA damage induced by fibres have been inconsistent, several indirect effects of DNA damage have been measured. For example, asbestos fibres have been shown to induce DNA repair synthesis (reviewed by Jaurand, 1991) and genetic responses to oxidant stress (Janssen et al., 1994). Tissue and species differences in these adaptation and repair pathways may account for discrepancies among different in-vitro model systems. These differences may also be apparent in human populations exposed to fibres.

Individual susceptibility to oxidant stress and altered DNA repair mechanisms. Individual adaptive responses to oxidant stress and the ability to repair damaged DNA are dependent on multiple exogenous and endogenous factors. Few experiments have attempted to manipulate these variables in animal or human model systems to test the hypothesis that free radicals play a role in fibre carcinogenesis. Only the following two studies using human cells *in vitro* have investigated genetic susceptibility to injury induced by asbestos fibres: fibroblasts from patients with a defect in DNA repair were found to be more sensitive to asbestos toxicity (Yang et al., 1984); and primary mesothelial cells obtained from individuals lacking glutathione S-transferase (GST) M1 were also more sensitive to asbestos toxicity (Pelin et al., 1995).

As described earlier, asbestos fibres have been shown to catalyse lipid peroxidation, generating organic hydroperoxides and other lipid-derived radicals that may damage DNA. One defence mechanism against organic hydroperoxides is reduction by GST. Approximately 50% of people carry a homozygous deletion of the GSTM1 isoenzyme. Pelin *et al.* (1995) explored the hypothesis that the *GSTM1* null genotype predisposes to asbestos-induced genetic damage and the development of malignant mesothelioma. No increase in the frequency of the *GSTM1* null genotype was found in a series of 44 patients with asbestos-related malignant mesothelioma. Mesothelial cell lines derived from people with the *GSTM1* null genotype showed increased toxicity in response to asbestos, but no increase in cytogenetic abnormalities. This study suggests that alterations in GSTM1 activity do not predispose to cytogenetic damage or malignant mesothelioma associated with asbestos exposure.

Rare inherited chromosomal instability syndromes exist that reflect abnormalities in various DNA repair pathways. Patients with xeroderma pigmentosum, ataxia telangiectasia or Bloom's syndrome have increased susceptibility to cancer induced by ultraviolet or ionizing radiation. Only one study has investigated the sensitivity of fibroblasts derived from patients with xeroderma pigmentosum to asbestos fibres *in vitro*. This study reported increased cytotoxicity in these cultures (Yang *et al.*, 1984). Additional experiments to confirm and expand this finding are required.

Hypothesis 2: Fibres interfere physically with mitosis
An important characteristic of natural and man-made fibres is their ability to induce aneuploidy, polyploidy and binucleate cells in a variety of cells cultured *in vitro* (reviewed by Jaurand, 1989, 1991). In contrast to the clastogenic effects of fibres, the aneuploidogenic effects of fibres can be demonstrated in a wide variety of cell types (with the exception of lymphocytes) derived from both rodents and humans. On the basis of these observations, it is hypothesized that the aneuploidogenic effects of fibres contribute to their carcinogenicity. The experimental observations that support this hypothesis will be discussed briefly.

Requirement for phagocytosis. For fibres to interfere physically with the mitotic apparatus, two requirements must be fulfilled. First, fibres must be internalized by the target cell population. With the exception of lymphocytes, most types of cells phagocytize fibres *in vitro*. This process may be facilitated by binding to specific membrane receptors; for example, crocidolite asbestos binds to the scavenger receptor of macrophages (Resnick *et al.*, 1993). Second, the target cell population must be proliferating; this requirement will be discussed in detail subsequently. While both of these requirements are achieved readily *in vitro*, it has not been established whether asbestos fibres are internalized by the target cells giving rise to bronchogenic carcinoma or malignant mesothelioma *in vivo*. Few measurements of asbestos fibre burdens have been made on pleural tissue; in two studies, predominantly short fibres were found, which did not correlate with the lung fibre burden (Sébastien *et al.*, 1980; Dodson *et al.*, 1990). Additional studies to localize fibres in the target cell populations of the pleura and bronchial epithelium are required to test this hypothesis *in vivo*.

Fate of internalized fibres. Previous morphological studies of cells exposed to asbestos fibres *in vitro*, then examined by light or electron microscopy, demonstrated direct physical interaction between fibres and the nucleus or chromosomes during mitosis (Wang *et al.*, 1987; Jaurand, 1991). More recent investigations demonstrate that internalized fibres are surrounded by a phagolysosomal membrane thereby preventing direct physical or chemical interaction with or adsorption to chromosomes (Jensen *et al.*, 1994). Both short and long crocidolite asbestos fibres were observed to accumulate in the perinuclear region of amphibian lung epithelial cells *in vitro*; short fibres showed saltatory transport along microtubules while long fibres remained stationary (Cole *et al.*, 1991). Approximately 10–20 fibres were internalized by these large epithelial cells *in vitro* without any adverse effects on viability or cell division (Ault *et al.*, 1995).

Mechanisms of interference with mitosis. Previous studies using light microscopy on fixed cells suggested physical interference of long fibres with the mitotic spindle and chromosomal segregation, which resulted in lagging or sticky chromosomes during anaphase and the subsequent generation of aneuploid daughter cells (Hesterberg & Barrett, 1985). A recent study using video-enhanced time-lapse microscopy of living amphibian lung epithelial cells during mitosis has extended these initial observations (Ault et al., 1995). In this cell type, a keratin microfilament cage surrounds the nucleus, and this architecture is maintained during mitosis. Long crocidolite asbestos fibres were observed to be trapped in this keratin microfilament cage protruding into the spindle region of mitotic cells. In most cases, collisions between the protruding asbestos fibre and chromosomes did not interfere with their orderly segregation; there was no evidence for sticking of chromosomes to fibres. In rare cases, a protruding fibre did impede chromosomal migration to the spindle pole or cause a chromosomal break. These investigators hypothesize that long fibres are more easily trapped in the keratin microfilament cage than short fibres and that they are more likely to protrude into the spindle region during mitosis. It is proposed that the keratin cage is disrupted during mitosis of mesothelial cells, allowing long fibres to interfere more readily with mitosis in this cell type (Ault et al., 1995).

In addition to physical interference with chromosomal segregation, fibres may also disrupt mitosis by other mechanisms. Exposure of human mesothelial cells to amosite fibres in vitro caused disruption of their cytoskeletal organization; this disruption could enhance the ability of fibres to interfere with the mitotic spindle (Somers et al., 1991). In addition to changes in chromosomal number, asbestos fibres also induce binucleate daughter cells in dividing cell populations in vitro. This effect is postulated to be the result of physical interference of fibres with the cleavage furrow (Kenne et al., 1986).

Are these mechanisms responsible for the cytogenetic alterations observed in experimental and human mesothelioma cell lines? Multiple non-random cytogenetic alterations are characteristic of experimental and human mesothelioma cell lines (reviewed by Barrett, 1994). Monosomy, polysomy and partial deletions are frequent, especially in human chromosomes 1, 3 and 5–9. The relationship between chromosomal alterations induced by fibres and carcinogenesis is supported by the cytogenetic alterations that accompany in-vitro transformation of SHE cells or rat pleural mesothelial cells exposed to fibres (reviewed by Walker et al., 1992). However, not all cell types are transformed by exposure to asbestos fibres in vitro, although they do show aneuploidy (reviewed by Jaurand, 1991). Additional mechanisms may contribute to the genetic alterations responsible for the development of bronchogenic carcinomas and malignant mesotheliomas associated with exposure to asbestos fibres. First, oxidative damage to DNA has recently been shown to alter the methylation pattern of adjacent cytosines. Altered gene methylation, as well as point mutations resulting from mispairing at sites of oxidized bases, may alter gene expression in malignant neoplasms (Weitzman et al., 1994). Second, the cytogenetic abnormalities characteristic of malignant mesotheliomas may result from genetic instability arising from loss of the G1 cell cycle checkpoint (Lane, 1992). Inactivation of normal p53 disrupts this G1 cell cycle checkpoint. Altered p53 function can result from deletion or point mutation of the *p53* tumour suppressor gene or binding to viral proteins such as the SV40 T-antigen. As summarized earlier, mutations in the *p53* tumour suppressor gene occur frequently in human bronchogenic carcinomas, although the pattern of *p53* mutations in patients who were also exposed to asbestos has not been investigated. Point mutations in *p53* are infrequent in human malignant mesotheliomas; in addition, no mutations in *MDM2*, which codes for a p53-binding protein, were found in human mesothelioma cell lines (Gerwin, 1994). However, over-expression of p53 has been demonstrated in up to one-half of human malignant mesotheliomas (Mayall et al., 1992; Ramael et al., 1992). Although the functional status of the G1 cell cycle checkpoint has not been studied in human mesothelial cell lines, inactivation of the G1 cell cycle checkpoint could be responsible for genetic instability and multiple chromosomal alterations observed in human malignant mesotheliomas (Moyer et al., 1994).

Unanswered questions. At least two mechanisms responsible for the genotoxic effects of fibres have been proposed: (i) indirect DNA damage mediated by free radicals; and (ii) direct physical interference with the mitotic apparatus. These mechanisms are supported by observations of the genotoxic effects of asbestos fibres using cell cultures in vitro. The following questions must be addressed in order to validate these proposed mechanisms:

- Are in-vitro genotoxicity assays relevant to fibre carcinogenesis?
- Are in-vitro doses relevant for in-vivo exposures?
- Can the genotoxic effects of fibres be assessed *in vivo*?

Hypothesis 3: Fibres stimulate the proliferation of target cells

Cell proliferation is required for the fixation of mutations induced by genotoxic agents. Clonal expansion of initiated or pre-neoplastic cell populations is accelerated in a growth-stimulatory environment. An increased rate of cell proliferation also leads to populations of cells with increased levels of spontaneous mutations (reviewed by Barrett, 1994). Cell proliferation is balanced by terminal differentiation and death of cells by apoptosis. As summarized below, exposure to asbestos fibres under certain conditions is accompanied by cell proliferation; the effects of fibres on terminal differentiation and apoptosis have not yet been explored.

Experimental observations in vitro and in vivo. At high levels of exposure *in vitro*, asbestos fibres are toxic to a variety of target cell populations. This toxicity is hypothesized to be mediated by the generation of ROS (reviewed by Mossman & Marsh, 1989) or the inhibition of cell proliferation by physical interference with mitosis (Hesterberg & Barrett, 1985). However, under some conditions *in vitro*, asbestos fibres induce proliferation of hamster tracheobronchial epithelial cells (Sesko *et al.*, 1990), rat lung fibroblasts (Lasky *et al.*, 1995) and rat pleural mesothelial cells (Heintz *et al.*, 1993). Cell proliferation in these same target cell populations in the lung and pleura has also been observed following intratracheal instillation (Dodson & Ford, 1985; Adamson *et al.*, 1993), inhalation (Brody *et al.*, 1989) or intraperitoneal injection of asbestos fibres (Moalli *et al.*, 1987; Friemann *et al.*, 1990). These early proliferative responses occur in cell populations where a malignant neoplasm may arise, that is, the bronchial epithelium, type II alveolar epithelial cells and mesothelial cells. However, proliferation of macrophages, interstitial fibroblasts and endothelial cells is also observed in these in-vivo models (Branchaud *et al.*, 1989; Brody *et al.*, 1989); these early proliferative reactions are associated with angiogenesis and fibrosis characteristic of a wound-healing response.

Pott (1979) has proposed a mechanistic hypothesis linking the biopersistence of fibres at the target tissue with the stimulation of cell proliferation; this hypothesis states that fibres must persist and stimulate a sufficient number of population doublings to give rise to a pre-neoplastic cell population (Pott, 1987). So far, this hypothesis has not been tested experimentally.

Mechanisms responsible for stimulation of cell proliferation. Experimental evidence has been obtained for four potential mechanisms of growth stimulation in response to asbestos fibres: (i) compensatory cell proliferation in response to toxicity; (ii) stimulation of intracellular signal transduction pathways; (iii) direct mitogenesis; and (iv) induction of growth factor and growth factor receptor expression.

The evidence for these mechanisms will be summarized briefly. These potential mechanisms require direct interaction with or phagocytosis of fibres by the target cells. Alternatively, an indirect mechanism leading to the stimulation of cell proliferation is through the release of cytokines and growth factors from inflammatory cells. This indirect mechanism does not require direct contact between fibres and the target cell population; this indirect pathway will be discussed subsequently.

Injury to target cells has been demonstrated in the parietal mesothelial lining after direct intraperitoneal injection of crocidolite asbestos fibres in mice (Moalli *et al.*, 1987). Localized damage to the alveolar epithelium has also been proposed to facilitate translocation of fibres and growth factors into the interstitium of the lungs following either inhalation (Brody *et al.*, 1989) or intratracheal instillation of asbestos fibres (Adamson & Bowden, 1988). Cell proliferation is triggered at these localized sites of injury induced by the deposition of asbestos fibres. In the parietal mesothelium, the morphology and kinetics of mesothelial cell proliferation resemble the healing response of this tissue as triggered by other types of injury (Moalli *et al.*, 1987).

A second mechanism leading to stimulation of cell proliferation by fibres is the triggering of intracellular signal transduction pathways. Several of these biochemical events are common to asbestos fibres and other tumour promoters such as phorbol esters. There is substantial experimental evidence for the action of asbestos fibres as a tumour promoter, especially for the tracheobronchial epithelium (Hoskins et al., 1991). For example, using the model of hamster tracheal epithelial cells exposed to fibres in vitro, B.T. Mossman and co-workers demonstrated the following effects of asbestos fibres: increased expression of ornithine decarboxylase (Marsh & Mossman, 1991); activation of protein kinase C (Perderiset et al., 1991); and hydrolysis of inositol phospholipids (Sesko et al., 1990). Some of these biochemical changes may be initiated by ROS (Marsh & Mossman, 1991).

A third mechanism leading to cell proliferation is direct mitogenesis in the absence of cell toxicity. Direct mitogenesis requires phagocytosis or the binding of the fibres to cell surface receptors, although this latter mechanism has not been explored in cells other than phagocytes. B.T. Mossman and co-workers have obtained evidence for the mitogenic effects of fibres in vitro based on induction of proto-oncogene expression. In contrast to soluble tumour promoters such as phorbol esters, asbestos fibres cause a prolonged expression of the proto-oncogenes c-fos and c-jun, which may be responsible for the persistent growth stimulation of target cells (Heintz et al., 1993).

Finally, asbestos fibres may also trigger cell proliferation by inducing the expression of growth factors and growth factor receptors, activating an autocrine growth-stimulatory pathway. Growth activation by this mechanism has been demonstrated by increased expression of PDGF-AA and its receptor by rat lung fibroblasts exposed to chrysotile asbestos in vitro (Lasky et al., 1995). The missing link between exposure to fibres and induction of gene expression is the identification of the mechanism responsible for the turning on of transcription factors that regulate specific genes. One potential mechanism is oxidant stress induced by generation of ROS leading to activation of the nuclear transcription factor κB (NF-κB) (Moyer et al., 1994).

Unanswered questions. Exploration of the effects of fibres on gene expression and cell proliferation is an emerging area of research; experiments must be designed to address the following question:

• What is the relationship between (i) the acute effects of fibres on gene expression and (ii) chronic, persistent proliferation of target cell populations?

Hypothesis 4: Fibres provoke a chronic inflammatory reaction leading to the prolonged release of ROS, cytokines and growth factors in the lungs
Macrophages are the initial target cells of fibres and other particulates that deposit in the lungs or pleural and peritoneal spaces. Phagocytosis of asbestos fibres is accompanied by the activation of macrophages, which results in the increased synthesis and secretion of ROS as well as a variety of chemical mediators and cytokines. These mediators amplify the local inflammatory reaction. Persistence of asbestos fibres in the interstitium of the lungs or in the subpleural connective tissue may lead to a sustained chronic inflammatory reaction accompanied by fibrosis (reviewed by Driscoll, 1993; Oberdörster, 1994). The mechanisms responsible for these inflammatory reactions will be reviewed briefly.

Recruitment of inflammatory cells. Migration of macrophages to sites of asbestos fibre deposition is mediated by activation of chemotactic factors. In the lungs, the chemotactic factor, C5a, has been shown to play an important role in the recruitment of macrophages (reviewed by Rom et al., 1991). Activated macrophages also synthesize arachidonic acid metabolites such as leukotriene B4, which are chemotactic for neutrophils. Activated macrophages also express cytokines, including interleukin-1 and tumour necrosis factor-α. These cytokines sustain and amplify the initial inflammatory response by triggering the release of additional chemotactic factors called chemokines (interleukin-8 and macrophage inflammatory protein 2) from macrophages, pulmonary epithelial cells (Driscoll, 1993) and mesothelial cells (Griffith et al., 1994).
Mediators released from inflammatory cells. Direct stimulation by phagocytosis of fibres (Donaldson et al., 1989, 1992) and indirect stimulation by cytokines triggers the synthesis and release of additional mediators. Neutrophils and macrophages release ROS as described above; these

may cause DNA and chromosomal damage. In addition, proteolytic and hydrolytic enzymes may be released, which damage basement membranes and connective tissue in the lungs. Growth factors for epithelial cells and fibroblasts, such as TGF-α and PDGF, are also released from activated macrophages. Release of these mediators is a nonspecific response to lung injury and inflammation; overall, there is a balance between these cytokines and growth factors, which restores the lung to its original function and structure. It is hypothesized that an imbalance between these cytokines and growth factors may contribute to the pathological effects of asbestos fibres, especially diffuse interstitial pulmonary fibrosis or asbestosis (Oberdörster, 1994).

Biological effects of inflammatory mediators. Unregulated or persistent release of these inflammatory mediators may lead to tissue injury, scarring by fibrosis and proliferation of epithelial and mesenchymal cells. The cytotoxic effects of asbestos fibres are amplified in the presence of neutrophils in in-vitro models (Kamp *et al.*, 1989; Kinnula *et al.*, 1995). Damage to the alveolar epithelial lining and basement membrane is especially dangerous because asbestos fibres and inflammatory mediators gain access to the interstitium of the lung (Rom *et al.*, 1991). The interrelationship between inflammation, lung injury and fibrosis has been established in two animal models. First, exposure of complement-deficient mice to chrysotile asbestos fibres resulted in reduced levels of macrophage accumulation in sites of fibre deposition and this attenuated the subsequent fibrotic reaction (McGavran *et al.*, 1989). Second, lung inflammation and fibrosis induced by inhalation of crocidolite asbestos fibres was decreased by polyethylene glycol conjugated to catalase, implicating ROS in lung injury produced by fibres (Mossman *et al.*, 1990).

Pulmonary fibrosis is frequently accompanied by proliferation of type II alveolar epithelial cells (Kuhn *et al.*, 1989). The relationship between sustained epithelial cell proliferation, diffuse interstitial pulmonary fibrosis and bronchogenic carcinoma is controversial.

Mesothelial cell proliferation, especially involving the visceral pleura, is also an early reaction to intratracheal instillation of asbestos fibres. However, at these early time points, no fibres have been observed in the visceral pleura and this proliferative response is hypothesized to be mediated indirectly by cytokines and growth factors released from activated interstitial macrophages (Adamson *et al.*, 1994). The relationship between chronic or persistent release of cytokines and growth factors, chronic mesothelial cell proliferation and the development of malignant mesothelioma remains to be tested.

Unanswered questions. In the lungs and pleural linings, chronic inflammation and fibrosis are common reactions following exposure to asbestos fibres. An important mechanistic question remains to be answered:

- What are the links between inflammation, fibrosis and cancer induced by fibres?

Hypothesis 5: Fibres act as co-carcinogens or carriers of chemical carcinogens to the target tissue
Cigarette smoke, asbestos and lung cancer. Epidemiological studies have confirmed that cigarette smoke plus asbestos fibres are multiplicative in the induction of bronchogenic carcinoma (Selikoff *et al.*, 1968; Saracci, 1977). Experiments using cell culture and animal models provide evidence that asbestos fibres enhance the delivery of multiple carcinogens in cigarette smoke to the bronchial epithelium and increase their metabolic activation. For example, cigarette smoking retards ciliary action and clearance of fibres and other particulates (McFadden *et al.*, 1986). Also, ROS in cigarette smoke or generated by iron-catalysed reactions in the presence of asbestos fibres enhance fibre uptake by tracheal epithelial cells (Hobson *et al.*, 1990). Both crocidolite and chrysotile asbestos also enhance cellular uptake and metabolic activation of benzo[*a*]pyrene in hamster tracheal epithelial cells (Mossman *et al.*, 1983). On the basis of these experimental observations, it is hypothesized that asbestos fibres and cigarette smoke act as co-carcinogens as summarized in Fig. 1.

As reviewed earlier, asbestos fibres may also contribute by additional mechanisms to the development of bronchogenic carcinoma, either alone or in combination with cigarette smoke. For example, asbestos and cigarette smoke act synergistically to damage DNA in a cell-free system; this

DNA damage has been shown to be mediated by the iron-catalysed generation of OH• (Jackson et al., 1987). In hamster tracheobronchial rings maintained in organ culture, asbestos fibres induce squamous metaplasia, a potentially reversible precursor lesion that may evolve into bronchogenic carcinoma (Mossman et al., 1977). Finally, asbestos fibres have been shown to stimulate the proliferation of tracheal epithelial cells and to mimic the biochemical effects of tumour promoters as discussed above.

Interactions between fibrous and non-fibrous dusts. People are commonly exposed to mixtures of fibrous and non-fibrous dusts, even in the ambient environment. Man-made fibres are also mixtures of fibrous and non-fibrous materials. The toxicity of particulates containing quartz is well recognized; however, less is known about the potential additive or synergistic effects between fibrous and less toxic non-fibrous dusts. Exposure of rodents to mixed dusts by inhalation resulted in increased transport of fibres across the visceral pleura and increased production of lung tumours and mesothelioma (reviewed by Oberdörster, 1994).

Unanswered questions. These experiments in rodents raise the following concern about exposure of humans to mixed fibrous and non-fibrous dusts:

- Does inhalation of fibres or mixed fibrous and non-fibrous dusts impair clearance in rats? Is this mechanism relevant for humans?

Caveats about fibres and carcinogenesis
The five mechanistic hypotheses above have been proposed for the role of asbestos fibres in the development of bronchogenic carcinoma and malignant mesothelioma. As summarized above and in the accompanying reviews, there is experimental evidence to support each of these mechanisms; however, there is not yet conclusive evidence for or against any of these mechanisms. Most of the current experimental data are derived from observations with asbestos fibres and a few types of man-made fibres. For these reasons, the following caveats must be considered:

Different types of fibres may act by different mechanisms. Experimental evidence has been presented for genetic and epigenetic roles of asbestos fibres in carcinogenesis. Different fibre types may show predominantly genetic or epigenetic effects. ROS are hypothesized to mediate some of the genetic and epigenetic effects of asbestos fibres. This mechanism must be evaluated critically for each type of natural and man-made fibre tested in in-vitro or in-vivo models. The problem of acute versus chronic or persistent effects must be addressed in appropriate animal model systems.

Different mechanisms may contribute to the development of bronchogenic carcinoma and malignant mesothelioma. The interaction between asbestos exposure and cigarette smoking in the development of bronchogenic carcinoma has

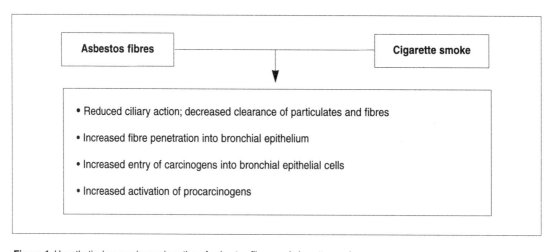

Figure 1. Hypothetical co-carcinogenic action of asbestos fibres and cigarette smoke.

been well documented. The experimental and epidemiological data support a role for asbestos fibres during the promotion stage of the development of bronchogenic carcinoma. Mechanisms have been proposed for the action of asbestos fibres as both initiators and promoters in the development of malignant mesothelioma.

Other potential mechanisms. The pathogenesis of malignant mesothelioma has not yet been established; therefore, it is important to explore all potential mechanisms. For example, Appel *et al.* (1988) demonstrated that asbestos fibres can transfect plasmid DNA into cells *in vitro*. Transfection of exogenous DNA can be mediated by other agents, including calcium phosphate precipitates, so the specificity of this mechanism for asbestos fibres was questioned. Recently, SV40-like DNA sequences were identified in 60% of human mesothelioma tissue samples, but not in adjacent lung tissue or in 47 other tumour samples studied. The origin of this viral DNA and the specificity for malignant mesothelioma tissue is unknown (Carbone *et al.*, 1994). Between 1954 and 1961, people ingested live oral polio vaccine that may have been contaminated with SV40 virus. Many people are exposed to related papovaviruses such as JC virus in childhood; this virus becomes latent and persists in the kidney. A related papovavirus, BK, has been found in brain tumours, Kaposi's sarcoma and in pancreatic islet-cell tumours (King, 1993). Asbestos fibres may transfect viral DNA into mesothelial cells. The presence of SV40-related DNA sequences and expression of the viral T-antigen would explain the high percentage of p53 immunoreactivity in human malignant mesothelioma tissue as discussed above (see *Molecular alterations in asbestos-related neoplasms*).

Alternatively, reactivation of a latent papovavirus infection in patients with malignant mesothelioma may be associated with an immunocompromised state. Alterations in cell-mediated immunity and natural killer cell activity have been described in patients with asbestosis (reviewed by Rom *et al.*, 1991). It is unknown whether altered immunosurveillance contributes to the development of asbestos-related neoplasms.

Fibres may contribute to carcinogenesis by multiple mechanisms. As discussed by Barrett (1994), fibres may act at multiple stages in neoplastic development. The co-carcinogenic effects of asbestos fibres and cigarette smoke in metabolic activation of chemical carcinogens have been summarized in Fig. 1. On the basis of molecular studies of lung cancers associated with cigarette smoking and asbestos exposure, it is proposed that the initiation of mutations in the early states of the development of bronchogenic carcinomas reflects the mutagenic activity of agents found in cigarette smoke (Husgafvel-Pursiainen *et al.*, 1993; Nuorva *et al.*, 1994). Both asbestos fibres and cigarette smoke then act as promoters to stimulate clonal proliferation of initiated cells. Pre-neoplastic epithelial cells in the lungs show histological evidence of altered differentiation (metaplasia) and altered proliferation (hyperplasia and dysplasia). Asbestos fibres have been shown to (i) induce metaplasia in tracheobronchial epithelium (Mossman *et al.*, 1977) and (ii) alter the temporal expression of proto-oncogenes that regulate proliferation of bronchial epithelial cells *in vitro* (Heintz *et al.*, 1993). Later stages in tumour development are characterized by additional genetic and cellular alterations leading to a locally invasive neoplasm followed by metastasis. The role of asbestos fibres in these later stages of tumour progression is unknown.

The proposed steps leading to the development of malignant mesothelioma are more speculative. Asbestos fibres may directly or indirectly injure and/or stimulate the proliferation of mesothelial cells. Fibres may be genotoxic to these proliferating cells. Alternatively, chronic stimulation of mesothelial cell proliferation may lead to the accumulation of spontaneous mutations that confer a proliferative advantage to pre-neoplastic cell populations. Fibres that persist in the interstitium or in the submesothelial connective tissue may trigger the chronic release of cytokines and growth factors from activated macrophages and thus lead to persistent stimulation of mesothelial cell growth. As in bronchogenic carcinoma, the role of asbestos fibres in the development of invasive and metastatic malignant mesothelioma is unknown.

Validation of mechanistic data in people exposed to asbestos or man-made fibres
The above proposed mechanistic hypotheses are based on experimental data derived from in-vitro models or in-vivo exposure of laboratory animals to

asbestos fibres. The endpoints of the in-vivo models have been validated by the production of the same spectrum of diseases observed in humans exposed to asbestos fibres, i.e. fibrosis, lung cancer and malignant mesothelioma (McClellan & Hesterberg, 1994). To use these mechanistic data in future evaluations of the carcinogenicity of natural and man-made fibres for humans, it is necessary to evaluate whether the same mechanisms are responsible for asbestos-related malignant diseases in both laboratory animals and humans. The following limited number of human studies have begun to investigate this issue:

Biomarkers in peripheral blood. Peripheral blood lymphocytes have been used as surrogates for lung cells to demonstrate an increased incidence of sister chromatid exchange in asbestos workers (Rom *et al.*, 1983). Workers exposed to quartz or asbestos fibres were also found to have increased levels of plasma lipid peroxides (Kamal *et al.*, 1989).

Bronchoalveolar lavage. Sampling of the bronchoalveolar compartment of the lungs in asbestos workers has confirmed persistent inflammation and the elevated release of ROS, cytokines, and growth factors from alveolar macrophages (Robinson *et al.*, 1986; Rom *et al.*, 1987; Zhang *et al.*, 1993). These changes are not entirely specific to exposure to asbestos fibres – they are also associated with other pneumoconioses and idiopathic pulmonary fibrosis (reviewed by Rom *et al.*, 1991).

Human lung and pleural biopsies. Biopsy or autopsy specimens of human lung and pleural tissues are an important resource with which to explore the cellular and molecular events in hyperplastic, dysplastic and neoplastic lesions associated with exposure to asbestos fibres. As reviewed by Gerwin (1994), this approach complements and confirms studies using cell lines derived from human bronchogenic carcinomas and malignant mesotheliomas. Few studies have been reported so far in patients with a history of asbestos exposure; additional studies are required to demonstrate specific cellular and molecular alterations in asbestos-related neoplasms.

Investigations using human tissue samples are essential to confirm whether the mechanisms identified in animal models also operate in humans. However, these studies must be designed carefully and evaluated critically. Direct sampling of preneoplastic target cell populations is an invasive procedure; however, selection of surrogate samples such as bronchoalveolar lavage fluid, pleural effusion or peripheral blood must be validated. Appropriate control populations must also be evaluated in order to asses the specificity of a potential biomarker for asbestos exposure and to control for confounding factors such as cigarette smoking. The long latent period between exposure to fibres and the onset of symptoms of disease complicates evaluation of causal relationships. Despite these potential problems and limitations, additional cellular and molecular investigations of asbestos-related diseases using human tissues are essential. Workers exposed to man-made fibres who develop pleural plaques or effusions should also be included in these investigations (McClellan *et al.*, 1992).

Acknowledgements
The research reported from the author's laboratory was supported by grants from the National Institute of Environmental Health Sciences (R01 ES03721 and R01 ES05712) and the American Cancer Society.

References
Adamson, I.Y.R. & Bowden, D.H. (1988) Relationship of alveolar epithelial injury and repair to induction of pulmonary fibrosis. *Am. J. Pathol.*, 130, 377–383

Adamson, I.Y.R., Bakowska, J. & Bowden, D.H. (1993) Mesothelial cell proliferation after instillation of long or short asbestos fibers in mouse lung. *Am. J. Pathol.*, 142, 1209–1216

Adamson, I.Y.R., Bakowska, J. & Bowden, D.H. (1994) Mesothelial cell proliferation: a nonspecific response to lung injury associated with fibrosis. *Am. J. Respir. Cell. Mol. Biol.*, 10, 253–258

Appel, J.D., Fasy, T.M., Kohtz, D.S., Kohtz, J.D. & Johnson, E.M. (1988) Asbestos fibers mediate transformation of monkey cells by exogenous plasmid DNA. *Proc. Natl Acad. Sci. USA*, 85, 7670–7674

Ault, J.G., Cole, R.W., Jensen, C.G., Jensen, L.C.W., Bachert, L.A. & Rieder, C.L. (1995) Behavior of crocidolite asbestos during mitosis in living vertebrate lung epithelial cells. *Cancer Res.*, 55, 792–798

Barrett, J.C. (1994) Cellular and molecular mechanisms of asbestos carcinogenicity: implications for biopersistence. *Environ. Health Perspectives*, 102 (Suppl. 5), 19–23

Barsky, S.H., Huang, S. & Bhuta, S. (1986) The extracellular matrix of pulmonary scar carcinomas is suggestive of a desmoplastic origin. *Am. J. Pathol.*, 124, 412–419

Barsky, S.H., Cameron, R., Osann, K.E., Tomita, D. & Holmes, E.C. (1994) Rising incidence of bronchioloalveolar lung carcinoma and its unique clinicopathologic features. *Cancer*, 73, 1163–1170

Becklake, M.R. (1994) Symptoms and pulmonary functions as measures of morbidity. *Ann. Occup. Hyg.*, 38, 569–580

Becklake, M.R. & Case, B.W. (1994) Fiber burden and asbestos-related lung disease: determinants of dose–response relationships. *Am. J. Respir. Crit. Care Med.*, 150, 1488–1492

Bignon, J. (1989) Mineral fibres in the non-occupational environment. In: Bignon, J., Peto, J. & Saracci, R., eds, *Non-Occupational Exposure to Mineral Fibres* (IARC Scientific Publications No. 90), Lyon, IARC, pp. 3–29

Boutin, C. & Rey, F. (1993) Thoracoscopy in pleural malignant mesothelioma: a prospective study of 188 consecutive patients. Part. I: Diagnosis. *Cancer*, 72, 389–393

Branchaud, R.M., Macdonald, J.L. & Kane, A.B. (1989) Induction of angiogenesis by intraperitoneal injection of asbestos fibres. *FASEB J.*, 3, 1747–1752

Brody, A.R., McGavran, P.D. & Overby, L.H. (1989) Brief inhalation of chrysotile asbestos induces rapid proliferation of bronchiolar–alveolar epithelial and interstitial cells. In: Bignon, J., Peto, J. & Saracci, R., eds, *Non-occupational Exposure to Mineral Fibres* (IARC Scientific Publications No. 90), Lyon, IARC, pp. 93–99

Cagle, P.T. (1995) Tumors of the lung (excluding lymphoid tumours). In: Thurlbeck, W.M. & Churg, A.M., eds, *Pathology of the Lung,* 2nd edn, New York, Thieme Medical Publishers, pp. 437–553

Carbone, M., Pass, H.I., Rizzo, P., Marinetti, M.R., DiMuzio, M., Mew, D.J.Y., Levine, A.S. & Procopio, A. (1994) Simian virus 40-like DNA sequences in human pleural mesothelioma. *Oncogene*, 9, 1781–1790

Case, B.W. (1994) Biological indicators of chrysotile exposure. *Ann. Occup. Hyg.*, 38, 503–518

Churg, A. (1994) Deposition and clearance of chrysotile asbestos. *Ann. Occup. Hyg.*, 38, 625–633

Churg, A.M. & Green, F.H.Y. (1995) Occupational lung disease. In: Thurlbeck, W.M. & Churg, A.M., eds, *Pathology of the Lung,* 2nd ed., New York, Thieme Medical Publishers, pp. 851–930

Cole, R.W., Ault, J.G., Hayden, J.H. & Rieder, C.O. (1991) Crocidolite asbestos fibers undergo size-dependent microtubule-mediated transport after endocytosis in vertebrate lung epithelial cells. *Cancer Res.*, 51, 4942–4947

Cote, R.J., Jhanwar, S.C., Novick, S. & Pellicer, A. (1991) Genetic alterations of the p53 gene are a feature of malignant mesotheliomas. *Cancer Res.*, 51, 5410–5416

Craighead, J.E., Abraham, J.L., Churg, A., Green, F.H.Y., Kleinerman, J., Pratt, P.C., Seemayer, T.A., Vallayathan, V. & Weill, H. (1982) The pathology of asbestos-associated disease of the lungs and pleural cavities: Diagnostic criteria and proposed grading schema. *Arch. Pathol. Lab. Med.*, 106, 544–596

Davis, J.M.G. (1992) Forum: The need for standardized testing procedures of all products capable of liberating respirable fibres: the example of materials based on cellulose. *Br. J. Ind. Med.*, 50, 187–190

Davis, J.M.G. (1994) The role of clearance and dissolution in determining the durability or biopersistence of mineral fibers. *Environ. Health Perspectives*, 102 (Suppl. 5), 113–117

De Vuyst, P. (1994) Biopersistence of respirable synthetic fibers and minerals: point of view of the chest physician. *Environ. Health Perspectives*, 102 (Suppl. 5), 7–9

Dodson, R.F. & Ford, J.O. (1985) Early response of the visceral pleura following asbestos exposure: an ultrastructural study. *J. Toxicol. Environ. Health*, 15, 673–686

Dodson, R.F., Williams, M.J., Corn, C.J., Brollo, A. & Bianchi, C. (1990) Asbestos content of lung tissue, lymph nodes and pleural plaques from former shipyard workers. *Am. Rev. Respir. Dis.*, 142, 843–847

Donaldson, K., Brown, G.M., Brown, D.M., Bolton, R.E. & Davis, J.M.G. (1989) Inflammation generating potential of long and short fibre amosite asbestos samples. *Br. J. Ind. Med.*, 46, 271–276

Donaldson, K., Li, X.Y., Dogra, S., Miller, B.G. & Brown, G.M. (1992) Asbestos-stimulated tumour necrosis factor release from alveolar macrophages depends on fibre length and opsonization. *J. Pathol.*, 168, 243–248

Dong, H., Buard, A., Renier, A., Levy, F., Saint-Etienne, L. & Jaurand, M.-C. (1994) Role of oxygen derivatives in the cytotoxicity and DNA damage produced by asbestos on rat pleural mesothelial cells in vivo. *Carcinogenesis*, 15, 1251–1255

Driscoll, K.E. (1993) In-vitro evaluation of mineral cytotoxicity and inflammatory activity. In: Guthrie, G.D. & Mossman, B.T., eds, *Reviews in Mineralogy*, Vol. 28, Chelsea, MI, Bookcrafters, pp. 489–511

Dunnigan, J. (1984) Biological effects of fibers: Stanton's hypothesis revisited. *Environ. Health Perspectives*, 57, 333–337

Eborn, S.K. & Aust, A.E. (1996) Effect of iron acquisition on induction of DNA single-strand breaks by erionite, a carcinogenic mineral fiber. *Arch. Biochem. Biophys.* 316, 507–514

Ellouk, S.A. & Jaurand, M.-C. (1994) Review of animal/*in vitro* data on biological effects of man-made fibers. *Environ. Health Perspectives*, 102 (Suppl. 2), 47–63

Friemann, J., Muller, K.M. & Pott, F. (1990) Mesothelial proliferation due to asbestos and man-made fibres. Experimental studies on rat omentum. *Pathol. Res. Pract.*, 186, 117–123

Fubini, B. (1993) The possible role of surface chemistry in the toxicity of inhaled fibers. In: Warheit, D.B., ed., *Fiber Toxicology*, San Diego, Academic Press, pp. 229–258

Gerwin, B.I. (1994) Asbestos and the mesothelial cell: a molecular trail to mitogenic stimuli and suppressor gene suspects. *Am. J. Respir. Cell. Mol. Biol.*, 11, 507–508

Goodglick, L.A. & Kane, A.B. (1990) Cytotoxicity of long and short crocidolite asbestos fibers *in vitro* and *in vivo*. *Cancer Res.*, 50, 5153–5163

Goodglick, L.A., Pietras, L.A. & Kane, A.B. (1989) Evaluation of the casual relationship between crocidolite asbestos-induced lipid peroxiation and toxicity to macrophages. *Am. Rev. Respir. Dis.*, 139, 1265–1273

Gordon, J.R.C. (1992) Medicolegal issues in mesothelioma and asbestos disease litigation. In: Henderson, D.W., Shilkin, K.B., Langlois, S.L.P., Whitaker, D., eds, *Malignant Mesothelioma*, New York, Hemisphere Building Corporation, pp. 328–350

Graceffa, P. & Weitzman, S.A. (1987) Asbestos catalyzes the formation of the 6-oxobenzo[*a*]pyrene radical from 6-hydroxybenzo[*a*]pyrene. *Arch. Biochem. Biophys.*, 257, 481–484

Griffith, D.E., Miller, E.J., Gray, L.D., Idell, S. & Johnson, A.R. (1994) Interleukin-1-mediated release of interleukin-8 by asbestos-stimulated human pleural mesothelial cells. *Am. J. Respir. Cell. Mol. Biol.*, 10, 245–252

Gulumian, M. & Kilroe-Smith, T.A. (1987) Crocidolite-induced lipid peroxidation in rat lung microsomes. *Environ. Res.*, 43, 267–273

Hardy, J.A. & Aust, A.E. (1995) The effect of iron binding on the ability of crocidolite asbestos to catalyze DNA single-strand breaks. *Carcinogenesis*, 16, 319–325

Harris, C.C. (1992) Tumour suppressor genes, multistage carcinogenesis and molecular epidemiology. *In:* Vainio, H., Magee, P., McGregor, D. & McMichael, A.J., eds, *Mechanisms of Carcinogenesis in Risk Identification* (IARC Scientific Publications No. 116), Lyon, IARC, pp. 67–88

Hei, T.K., Piao, C.Q., He, Z.Y., Vannais, D. & Waldren, C.A. (1992) Chrysotile fiber is a strong mutagen in mammalian cells. *Cancer Res.*, 52, 6305–6309

Heintz, N.H., Janssen, Y.M. & Mossman, B.T. (1993) Persistent induction of c-fos and c-jun expression by asbestos. *Proc. Natl Acad. Sci. USA*, 90, 3299–3303

Henderson, D.W., Shilkin, K.B., Whitaker, D., Attwood, H.D., Constance I.J., Steele, R.H. & Leppard, P.J. (1992) The pathology of malignant mesothelioma, including immunohistology and ultrastructure. In: Henderson, D.W., Shilkin, K.B., Langlois, S.L.P. & Whitaker, D., eds, *Malignant Mesothelioma*, New York, Hemisphere Publishing Corporation, pp. 69–139

Hesterberg, T.W. & Barrett, J.C. (1985) Induction by asbestos fibers of anaphase abnormalities: mechanism for aneuploidy induction and possible carcinogenesis. *Carcinogenesis*, 6, 473–475

Hesterberg, T.W. & Barrett, J.C. (1987) Dependence of asbestos- and mineral dust-induced transformation of mammalian cells in culture on fiber dimension. *Cancer Res.*, 47, 1681–1686

Hesterberg, T.W., Miller, W.C., Mast, R., McConnell, E.E., Bernstein, D.M. & Anderson, R. (1994) Relationship between lung biopersistence and biological effects of man-made vitreous fibers after chronic inhalation in rats. *Environ. Health Perspectives*, 102 (Suppl. 5), 133–137

Hillerdall, G. (1994) Pleural plaques and risk for bronchial carcinoma and mesothelioma. A prospective study. *Chest*, 105, 144–150

Hobson, J., Wright, J.L. & Churg, A. (1990) Active oxygen species mediate asbestos fibre uptake by tracheal epithelial cells. *FASEB J.*, 4, 3134–3139

Hollstein, M., Sidransky, P., Vogelstein, B. & Harris, C.C. (1991) p53 Mutations in human cancers. *Science*, 253, 49–53

Hoskins, J.A., Brown, R.C. & Evans, C.E. (1991) Promoting effects of fibres. Fibres and the second messenger pathways. In: Brown, R.C., Hoskins, J.A. & Johnson, N.F., eds, *Mechanisms in Fibre Carcinogenesis* (NATO ASI Series), New York, Plenum Press, pp. 459–467

Husgafvel-Pursiainen, K., Hackmann, P., Ridanpaa, M., Anttila, S., Karjalinen, A., Partanien, T., Taikina-Aho, O.,

Heikkila, L. & Vainio, H. (1993) K-ras mutations in human adenocarcinoma of the lung: association with smoking and occupational exposure to asbestos. *Int. J. Cancer*, 53, 250–256

IARC (1977) *IARC Monographs on the Evaluation of Carcinogenic Risk of Chemicals to Man*, Vol. 14, *Asbestos*, Lyon, pp. 11–106

IARC (1987a) *IARC Monographs on the Evaluation of Carcinogenic Risk of Chemicals to Humans*, Vol. 42, *Silica and Some Silicates*, Lyon, pp. 33–249

IARC (1987b) *IARC Monographs on the Evaluation of Carcinogenic Risks to Humans*, Suppl. 7, *Overall Evaluations of Carcinogenicity: An Updating of IARC Monographs Volumes 1 to 42*, Lyon, pp. 106–117, 203

IARC (1988) *IARC Monographs on the Evaluation of Carcinogenic Risks to Humans*, Vol. 43, *Man-made Mineral Fibres and Radon*, Lyon, pp. 39–171

Jackson, J.H., Schraufstatter, I.U., Hyslop, P.A., Vosbeck., K., Sauerheber, R., Weitzman, S.A. & Cochrane, C.G. (1987) Role of oxidants in DNA damage: hydroxyl radical mediates the synergistic DNA damaging effects of asbestos and cigarette smoke. *J. Clin. Invest.*, 80, 1090–1095

Janssen, Y.M.W., Marsh, J.P., Absher, M.P., Gabrielson, E., Borm, P.J.A., Driscoll, K. & Mossman, B.T. (1994) Oxidant stress responses in human pleural mesothelial cells exposed to asbestos. *Am. J. Respir. Crit. Care Med.*, 149, 795–802

Jaurand, M.-C. (1989) Particulate-state carcinogenesis: a survey of recent studies on the mechanisms of action of fibres. In: Bignon, J., Peto, J. & Saracci, R., eds, *Non-Occupational Exposure to Mineral Fibres* (IARC Scientific Publications No. 90), Lyon, IARC, pp. 54–73

Jaurand, M.-C. (1991) Mechanisms of fibre genotoxicity. In: Brown, R.C., Hoskins, J.A. & Johnson, N.F., eds, *Mechanisms in Fibre Carcinogenesis*, New York, Plenum Press, pp. 287–307

Jensen, C.G., Jensen, L.C.W., Ault, J.G., Osorio, G., Cole, R. & Rieder, C.C. (1994) Time-lapse video light microscopic and electron microscopic observations of vertebrate epithelial cells exposed to crocidolite asbestos. In: Davis, J.M.G. & Jaurand, M.-C., eds, *Cellular and Molecular Effects of Mineral and Synthetic Dusts and Fibres*, Berlin, Springer-Verlag, pp. 63–78

Kamal, A.-A.M., Gomaa, A., Khafif, M.E. & Hammad, A.S. (1989) Plasma lipid peroxides among workers exposed to silica or asbestos dusts. *Environ. Res.*, 49, 173–180

Kamp, D.W., Dunne, M., Weitzman, S.A. & Dunn, M.M. (1989) The interaction of asbestos and neutrophils injures cultured human pulmonary epithelial cells: role of hydrogen peroxide. *J. Lab. Clin. Med.*, 114, 604–612

Kamp, D.W., Graceffa, P., Pryor, W.A. & Weitzman, S.A. (1992) The role of free radicals in asbestos-induced diseases. *Free Rad. Biol. Med.*, 12, 293–315

Kenne, K., Ljungquist, S. & Ringertz, N.R. (1986) Effects of asbestos fibers on cell division, cell survival, and formation of thioguanine-resistant mutants in Chinese hamster ovary cells. *Environ. Res.*, 39, 448–464

King, N.W., Jr (1993) Simian virus 40 infection. In: Jones, T.C., Mohr, U. & Hunt, R.D., eds, *Nonhuman Primates I*, Berlin, Springer-Verlag, pp. 37–42

Kinnula, V.L., Everitt, J.I., Magum, J.B., Chang, L.-Y. & Crapo, J.D. (1992) Antioxidant defense mechanisms in cultured pleural mesothelial cells. *Am. J. Respir. Cell. Mol. Biol.*, 7, 94–103

Kinnula, V.L., Raivio, K.O., Linnainmaa, K., Ekman, A. & Klockars, M. (1995) Neutrophil and asbestos fiber induced cytotoxicity in cultured human mesothelial and bronchial epithelial cells. *Free Rad. Biol. Med.*, 18, 391–400

Kuhn, C., III, Boldt, J., King, T.E., Jr, Crouch, E., Vartio, T. & McDonald, J.A. (1989) An immunohistochemical study of architectural remodeling and connective tissue synthesis in pulmonary fibrosis. *Am. Rev. Respir. Dis.*, 140, 1693–1703

Lane, D.P. (1992) p53, Guardian of the genome. *Nature*, 358, 15–16

Lasky, J.A., Coin, P.G., Lindroos, P.M., Ostrowski, L.E., Brody, A.R. & Bonner, J.C. (1995) Chrysotile asbestos stimulates platelet-derived growth factor-AA production by rat lung fibroblasts *in vitro*: evidence for an autocrine loop. *Am. J. Respir. Cell. Mol. Biol.*, 12, 162–170

Leanderson, P., Soderkvist, P., Tagesson, C. & Axelson, O. (1988) Formation of 8-hydroxydeoxyguanosine by asbestos and man-made mineral fibres. *Br. J. Ind. Med.*, 45, 309–311

Leineweber, J.P. (1980) Dust chemistry and physics: mineral and vitreous fibres. In: Wagner, J.C., ed., *Biological Effects of Mineral Fibres*, Vol. 2 (IARC Scientific Publications No. 30), Lyon, IARC, pp. 881–900

Lippmann, M. (1994) Letter to the editor. Workshop on the health risk associated with chrysotile asbestos: a brief review. *Ann. Occup. Hyg.*, 38, 639–642

MacDonald, J.L. & Kane, A.B. (1986) Identification of asbestos fibers within single cells. *Lab. Invest.*, 55, 177–185

Marsh, J.P. & Mossman, B.T. (1991) Role of asbestos and active oxygen species in activation and expression of ornithine decarboxylase in hamster tracheal epithelial cells. *Cancer Res.*, 51, 167–173

Mast, R.W., Hesterberg, T.W., Glass, L.R., McConnell, E.E., Anderson, R. & Bernstein, D.M. (1994) Chronic inhalation and biopersistence of refractory ceramic fiber in rats and hamsters. *Environ. Health Perspectives*, 102, 207–209

Mayall, F.G., Goddard, H. & Gibbs, A.R. (1992) p53 Immuno-staining in the distinction between benign and malignant mesothelial proliferations using formalin-fixed paraffin sections. *J. Pathol.*, 168, 377–381

McClellan, R.O. & Hesterberg, T.W. (1994) Role of biopersistence in the pathogenicity of man-made fibers and methods for evaluating biopersistence: a summary of two round-table discussions. *Environ. Health Perspectives*, 102 (Suppl. 5), 277–283

McClellan, R.O., Miller, F.J., Hesterberg, T.W., Warheit, D.B., Bunn, W.B., Kane, A.B., Lippmann, M., Mast, R.W., McConnell, E.E. & Reinhardt, C.F. (1992) Approaches to evaluating the toxicity and carcinogenicity of man-made fibers: summary of a workshop held November 11–13, 1991, Durham, North Carolina. *Reg. Toxicol. Pharmacol.*, 16, 321–364

McFadden, D., Wright, J.L., Wiggs, B. & Churg, A. (1986) Smoking inhibits asbestos clearance. *Am. Rev. Respir. Dis.*, 133, 372–374

McGavran, P.D., Butterick, C.J. & Brody, A.R. (1989) Tritiated thymidine incorporation and the development of an interstitial lesion in the bronchiolar–alveolar regions of the lungs of normal and complement deficient mice after inhalation of chrysotile asbestos. *J. Environ. Pathol. Toxicol. Oncol.*, 9, 377–391

Metcalf, R.A., Welsh, J.A., Bennett, W.P., Seddon, M.B., Lehman, T.A., Pelin, K., Linnainmaa, K., Tammilehto, L., Mattson, K., Gerwin, B.I. & Harris, C.C. (1992) p53 and Kirsten-ras mutations in human mesothelioma cell lines. *Cancer Res.*, 52, 2610–2615

Meurman, L.O., Pukkala, E. & Hakama, M. (1994) Incidence of cancer among anthophyllite asbestos miners in Finland. *Occup. Environ. Med.*, 51, 421–425

Moalli, P.A., Macdonald, J.L., Goodglick, L.A. & Kane, A.B. (1987) Acute injury and regeneration of the mesothelium in response to asbestos fibers. *Am. J. Pathol.*, 128, 426–445

Mollo, F., Piolatto, G., Bellis, D., Adrion, A., Delsedime, L., Bernardi, P., Pira, E. & Ardissone, F. (1990) Asbestos exposure and histologic cell types of lung cancer in surgical and autopsy series. *Int. J. Cancer*, 46, 576–580

Monchaux, G., Bignon, J., Jaurand, M.-C., Lafuma, J., Sébastien, P., Masse, R. & Goni, R. (1981) Mesotheliomas in rats following inoculation with acid-leached chrysotile asbestos and other mineral fibres. *Carcinogenesis*, 2, 220–236

Morocz, I.A., Schmitter, D., Lauber, B. & Stahel, R.A. (1994) Autocrine stimulation of a human lung mesothelioma cell line is mediated through the transforming growth factor α/epidermal growth factor receptor mitogenic pathway. *Br. J. Cancer*, 70, 850–856

Mossman, B.T. & Marsh, J.P. (1989) Evidence supporting a role for active oxygen species in asbestos-induced toxicity and lung disease. *Environ. Health Perspectives*, 81, 91–94

Mossman, B.T., Kessler, J.B., Ley, B.N. & Craighead, J.E. (1977) Interaction of crocidolite asbestos with hamster respiratory mucosa in organ culture. *Lab. Invest.*, 36, 131–139

Mossman, B.T., Eastman, A., Landesman, J.M. & Bresnick, E. (1983) Effects of crocidolite and chrysotile asbestos on cellular uptake and metabolism of benzo[a]pyrene in hamster tracheal epithelial cells. *Environ. Health Perspectives*, 51, 331–335

Mossman, B.T., Marsh, J.P., Sesko, A., Hill, S., Shatos, M.A., Doherty, J., Petruska, J., Adler, K.B., Hemenway, D., Mickey, R., Vacek, P. & Kagan, E. (1990) Inhibition of lung injury, inflammation, and interstitial pulmonary fibrosis by polyethylene glycol-conjugated catalase in a rapid inhalation model of asbestosis. *Am. Rev. Respir. Dis.*, 41, 1266–1271

Moyer, V.D., Cistulli, C.A., Vaslet, C.A. & Kane, A.B. (1994) Oxygen radicals and asbestos carcinogenesis. *Environ. Health Perspectives*, 102 (Suppl. 5), 131–136

Musk, A.W. & Christmas, T.I. (1992) The clinical diagnosis of malignant mesothelioma. In: Henderson, D.W., Shilkin, K.B., Langlois, S.L.P. & Whitaker, D., eds, *Malignant Mesothelioma*, New York, Hemisphere Publishing Corporation, pp. 253–258

Nuorva, K., Makitaro, R., Huhti, E., Kamel, D., Vahakangas, K., Bloigu, R., Soini, Y. & Paakko, P. (1994) p53 Protein accumulation in lung carcinomas of patients exposed to asbestos and tobacco smoke. *Am. J. Respir. Crit. Care Med.*, 150, 528–533

Oberdörster, G. (1994) Macrophage-associated responses to chrysotile. *Ann. Occup. Hyg.*, 38, 601–615

Ohshima, H. & Bartsch, H. (1994) Chronic infections and inflammatory processes as cancer risk factors: possible role of nitric oxide in carcinogenesis. *Mutat. Res.*, 305, 253–264

Pelin, K., Hirvonen, A., Taavitsainen, M. & Linnainmaa, K. (1995) Cytogenetic response to asbestos fibers in cultured human primary mesothelial cells from 10 different donors. *Mutat. Res.*, 334, 225–233

Perderiset, J., Marsh, J.P. & Mossman, B.T. (1991) Activation of protein kinase C by crocidolite asbestos in hamster tracheal epithelial cells. *Carcinogenesis*, 12, 1499–1502

Pott, F. (1987) Problems in defining carcinogenic fibres. *Ann. Occup. Hyg.*, 31, 799–802

Pott, F. & Friedrichs, K.H. (1972) Tumours in rats after intraperitoneal injection of asbestos dusts. *Naturwissenschaften*, 59, 318–320

Ramael, M., Lemmens, G., Eerdekens, C., Buysse, C., Deblier, I., Jacobs, W. & Van Marck, E. (1992) Immunoreactivity for p53 protein in malignant mesothelioma and non-neoplastic mesothelium. *J. Pathol.*, 168, 371–375

Resnick, D., Freedman, N.J., Xu, S. & Kreiger, M. (1993) Secreted extracellular domains of macrophage scavenger receptors form elongated trimers which specifically bind crocidolite asbestos. *J. Biol. Chem.*, 268, 3538–3545

Robinson, B.W., Rose, A.H., James, A., Whitaker, D. & Musk, A.W. (1986) Alveolitis of pulmonary asbestosis: bronchoalveolar lavage studies in crocidolite- and chrysotile-exposed individuals. *Chest*, 90, 396–402

Rom, W.N., Livingston, G.K., Casey, K.R., Wood, S.D., Egger, M.J., Chiu, G.L. & Jerominski, L. (1983) Sister chromatid exchange frequency in asbestos workers. *J. Natl Cancer Inst.*, 70, 45–48

Rom, W.N., Bitterman, P.B., Rennard, S.I., Cantin, A. & Crystal, R.G. (1987) Characterization of the lower respiratory tract inflammation of non smoking individuals with interstitial lung disease associated with chronic inhalation of inorganic dusts. *Am. Rev. Respir. Dis.*, 136, 1429–1434

Rom, W.N., Travis, W.D. & Brody, A.R. (1991) Cellular and molecular basis of asbestos-related diseases. *Am. Rev. Respir. Dis.*, 143, 408–422

Saracci, R. (1977) Asbestos and lung cancer: an analysis of the epidemiological evidence on the asbestos smoking interaction. *Int. J. Cancer*, 20, 323–331

Schauer, I.E., Siriwardana, S., Langan, T.A. & Sclafani, R.A. (1994) Cyclin D1 overexpression vs. retinoblastoma inactivation: implications for growth control evasion in non-small cell and small cell lung cancer. *Proc. Natl Acad. Sci. USA*, 91, 7827–7831

Scheule, R.K. & Holian, A. (1989) IgG specifically enhances chrysotile asbestos-stimulated superoxide anion production by the alveolar macrophage. *Am. J. Respir. Cell Mol. Biol.*, 1, 313–318

Schwartz, D.A. (1991) New developments in asbestos-induced pleural disease. *Chest*, 99, 191–198

Sébastien, P. (1994) Biopersistence of man-made vitreous silicate fibers in the human lung. *Environ. Health Perspectives*, 102 (Suppl. 5), 225–228

Sébastien, P., Janson, X., Gaudichet, A., Hirsch, A. & Bignon, J. (1980) Asbestos retention in human respiratory tissues: comparative measures in lung parenchyma and in parietal pleura. In: Wagner, J.C., ed., *Biological Effects of Mineral Fibres* (IARC Scientific Publications No. 30), Lyon, IARC, pp. 237–246

Selikoff, I.J., Hammond, E.C. & Churg, J. (1968) Asbestos exposure, smoking, and neoplasia. *J. Am. Med. Assoc.*, 204, 106–112

Sesko, A., Cabot, M. & Mossman, B. (1990) Hydrolysis of inositol phospholipids precedees cellular proliferation in asbestos-stimulated tracheobronchial epithelial cells. *Proc. Natl Acad. Sci. USA*, 87, 7385–7389

Somers, A.N.A., Mason, E.A., Gerwin, B.I., Harris, C.C. & Lechner, J.F. (1991) Effects of amosite asbestos fibers on the filaments present in the cytoskeleton of primary human mesothelial cells. In: Brown, R.C., Hoskins, J.A. & Johnson, N.F., eds, *Mechanisms in Fibre Carcinogenesis*, New York, Plenum Press, pp. 481–490

Stanton, M.F. & Wrench, C. (1972) Mechanisms of mesothelioma induction with asbestos and fibrous glass. *J. Natl Cancer Inst.*, 48, 797–821

Stanton, M.F., Layard, M., Tegeris, A., Miller, E., May, M., Morgan, E. & Smith, A. (1981) Relation of particle dimension to carcinogenicity in amphibole asbestoses and other fibrous minerals. *J. Natl Cancer Inst.*, 67, 965–975

Sussman, R.G., Cohen, B.S. & Lippmann, M. (1991) Asbestos fiber deposition in a human tracheobronchial cast. II. Empirical model. *Inhal. Toxicol.*, 3, 161–179

Takeuchi, T. & Morimoto, K. (1994) Crocidolite asbestos increased 8-hydroxydeoxyguanosine levels in cellular DNA of a human promyelocytic leukemia cell line, HL60. *Carcinogenesis*, 15, 635–639

Tiainen, M., Kere, J., Tammilehto, L., Mattson, K. & Knuutila, S. (1992) Abnormalities of chromosomes 7 and 22 in human malignant pleural mesotheliomas: correlation between Southern blot and cytogenetic analyses. *Genes Chromosomes Cancer*, 4, 176–182

Vainio, H., Magee, P., McGregor, D. & McMichael, A.J., eds (1992) *Mechanisms of Carcinogenesis in Risk Identification* (IARC Scientific Publications No. 116), Lyon, IARC, pp. 9–54

Van der Meeren, A., Seddon, M.B., Betsholtz, C.A., Lechner, J.F. & Gerwin, B.I. (1993) Tumorigenic conversion of human mesothelial cells as a consequence of platelet-derived growth factor – A chain overexpression. *Am. J. Respir. Cell Mol. Biol.*, 8, 214–221

Viallet, J. & Minna, J.D. (1990) Dominant oncogenes and tumor suppressor genes in the pathogenesis of lung cancer. *Am. J. Respir. Cell Mol. Biol.*, 2, 225–232

Walker, C., Everitt, J. & Barrett, J.C. (1992) Possible cellular and molecular mechanisms for asbestos carcinogenicity. *Am. J. Ind. Med.*, 21, 253–273

Wang, N.S., Jaurand, M.-C., Magne, L., Kheuang, L., Pinchon, M.C. & Bignon, J. (1987) The interactions between asbestos fibers and metaphase chromosomes of rat pleural mesothelial cells in culture. *Am. J. Pathol.*, 126, 343–349

Warheit, D.B., Hartsky, M.A., McHugh, T.A. & Kellar, K.A. (1994) Biopersistence of inhaled organic and inorganic fibers in the lungs of rats. *Environ. Health Perspectives*, 102 (Suppl. 5), 151–157

Weitzman, S.A., Turk, P.W., Milkowski, D.H & Kozlowksi, K. (1994) Free radical adducts induce alterations in DNA cytosine methylation. *Proc. Natl Acad. Sci. USA*, 91, 1261–1264

Westra, W.H., Slebos, R.J.C., Offerhaus, G.J.A., Goodman, S.N., Evers, S.G., Kensler, T.W., Askin, F.B., Rodenhuis, S. & Hruban, R.H. (1993) K-*ras* oncogene activation in lung adenocarcinomas from former smokers. *Cancer*, 72, 432–438

Wheeler, C.S. (1990) Exposure to man-made mineral fibers: a summary of current animal data. *Toxicol. Ind. Health*, 6, 293–307

Yang, L.L., Kouri, E. & Curren, R.D. (1984) Xeroderma pigmentosum fibroblasts are more sensitive to asbestos fibers than are normal human fibroblasts. *Carcinogenesis*, 5, 291–294

Yu, C.P., Zhang, L., Oberdörster, G., Mast, R.W., Glass, L.R. & Utell, M.J. (1994) Clearance of refractory ceramic fibers (RCF) from the rat lung: development of a model. *Environ. Res.*, 65, 243–253

Zhang, Y., Lee, T.C., Guillemin, B., Yu, M.-C. & Rom, W.N. (1993) Enhanced IL-1β and tumor necrosis factor-α release and messenger RNA expression in macrophages from idiopathic pulmonary fibrosis or after asbestos exposure. *J. Immunol.*, 150, 4188–4196

Corresponding author
A.B. Kane
Department of Pathology and Laboratory Medicine, Brown University, Providence, RI 02912, USA

Use of physico-chemical and cell-free assays to evaluate the potential carcinogenicity of fibres

B. Fubini

Introduction

Physico-chemical characteristics of fibres determine biological responses

The adverse health effects of inhaled fibres depend upon the physico-chemical properties of the fibres (e.g. physical dimensions, solubility and surface chemistry). To evaluate the potential carcinogenicity of fibres, these physico-chemical characteristics should be investigated (McClellan et al., 1992).

An inhaled fibre interacts with biological matter in several different ways (Fig. 1), and the relevant reactions are governed by the surface chemistry of the fibre, which varies from one type of fibre to another (reviewed by Fubini, 1993). Surface properties of the fibre determine the following:

- Translocation of the fibre into the various biological compartments and its fate *in vivo* – this may be mediated by the hydrophilicity of the fibre surface, the presence of various adsorbed substances, the relative cell–fibre adhesion and any propensity for membranolysis.
- The activation of cells – this is related to the surface composition and charge of the fibre.
- Internalization – whether a fibre is engulfed by a cell depends not only on the shape of the fibre but also on its chemical composition at the surface. For example, iron-derived reactive oxygen/nitrogen species (ROS) mediate asbestos fibre uptake by tracheal epithelial cells (Churg, 1994); also, silica particles that have been heated are retained in lung tissue and in alveolar macrophages more than unheated ones (Hemenway et al., 1993). Figure 2 illustrates the role of surface properties at different stages of phagocytosis by alveolar macrophages.
- The potential for free radical release and consequent DNA damage (reviewed by Hardy & Aust, 1995a).
- The duration of biopersistence and clearance.

Figure 1 also indicates the fibre–cell interactions involved in some of the mechanisms hypothesized for mineral fibre carcinogenesis.

The physico-chemical properties thought to be relevant to fibre carcinogenicity

Form. The 'form' of a particle comprises its geometrical shape, crystallinity and micromorphology. Stanton et al. (1977) proposed that the carcinogenic potential of a fibre was associated with the length and aspect ratio of the fibre. This hypothesis, known as the Stanton hypothesis, has been discussed in recent years (Dunnigan, 1984; Jaurand, 1989; Goodglick & Kane, 1990). Several recent studies show evidence of a different biological response to 'long' and 'short' fibres (Adamson et al., 1993; Gilmour et al., 1995; Hill et al., 1995); however, it is acknowledged that form alone is not sufficient to explain the biological responses elicited. It is now considered that crystallinity and the type of exposed crystal faces determine reactivity and solubility. Also, micromorphology (i.e. surface irregularities, indenting, steps, kinks and edges) affect reactivity, solubility and, consequently, biopersistence (Hochella, 1993).

Chemical composition. The chemical composition of a fibre determines the following:

- its adsorption of exogenous materials (the fibre may act as a carrier of chemical carcinogens);
- its adsorption of endogenous materials, which may coat the fibre and inactivate or opsonize it;
- its potential for free radical generation at the solid–liquid interface, which may cause ROS formation, DNA damage and lipid peroxidation;
- its potential for the mobilization of metal ions by endogenous chelators;
- its propensity for selective leaching, and hence solubility and biopersistence.

(a) The fibre adsorbs xenobiotics prior to inhalation

(b) The fibre releases free radicals causing DNA damage

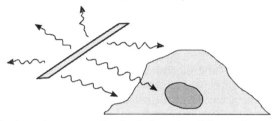

(c) The fibre adsorbs endogenous molecules in the extracellular medium or in the cytoplasm

(d) The fibre contacts directly with the target cells

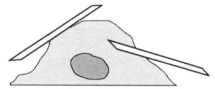

(e) The fibre is taken up by alveolar macrophages by phagocytosis, causing radical reaction within the phagolysosome and the prolonged release of cytokines and growth factors

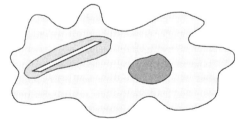

Figure 1. Interactions between fibres and cells that are thought to be involved in some of the mechanisms hypothesized for mineral fibre carcinogenesis. (a) A fibre may adsorb exogenous matter, prior to contact with a biological system; the fibre may then act as a carrier of toxic xenobiotics. (b) A fibre may release free radicals that damage DNA. (c) A fibre may adsorb biomolecules from body fluids, altering its properties in a manner deleterious to the target cell. (d) A fibre may contact directly with the mitotic spindle of the target cell, interfering with cell division. (e) A fibre may be taken up by alveolar macrophages by phagocytosis, leading to processes illustrated in Fig. 2.

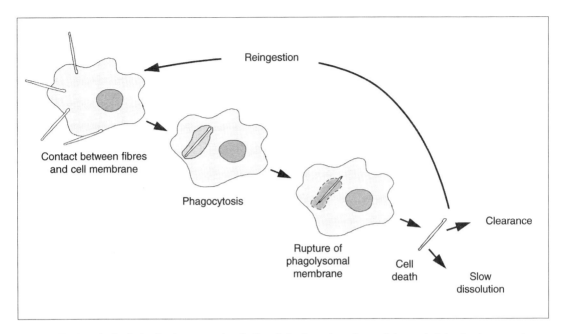

Figure 2. The hypothesized role of surface properties of a fibre during the various stages of phagocytosis by alveolar macrophages. Activation occurs upon contact with the cell membrane, and uptake into the cell is mediated by factors released by the fibre (e.g. reactive oxygen/nitrogen species or ROS). After phagocytosis, the fibre is in direct contact with the phagolysosomal contents, with consequent redox and radical reactions that may result in phagolysomal membrane rupture and death of the cell. Fibre reingestion may then occur. This cycle, a repeated series of unsuccessful phagocytosis events, generates a continuous release of ROS of endogenous and exogenous origins.

Solubility. Mineral fibres are poorly soluble materials. A wide range of solubility exists for synthetic mineral fibres, with many existing and newly developed fibres being highly soluble in the lung; those richer in alkaline oxides rather than alkaline earths and alumina are the most soluble. 'Solubility' *in vivo* is governed by several factors (see below), among which is solubility in water. In aqueous solutions, solubility may either be increased by specific reactions between solutes and some fibre component (e.g. endogenous chelators reacting with metal ions) or decreased by strong adsorption of molecules at the fibre surface.

Physico-chemical characterization of fibres
Fibre form
Dimensions, size distribution and exposed surface.
Fibre dimensions are given in length and diameter. Size distribution (the proportion of fibres in a sample within particular diameter–length classes) is a convenient way of describing a fibre and is usually evaluated by means of optical or transmission electron microscopy (TEM) (Johnson *et al.*, 1992); light-diffraction techniques are inappropriate when dealing with non-isometric particles. However, a 'geometrical' method such as this can not take into account the complex indenting, steps and edges at the atomic surface level of most fibres. However, the extent of this true or 'exposed' surface can be measured by means of the physical adsorption of nitrogen at low temperature, usually employing the Brunauer–Emmet–Teller (BET) method, and the difference between the geometrical surface area, calculated on the basis of fibre dimensions and distribution, and the exposed surface area represents a sort of fractal development of the surface, i.e. the non-smoothness of the surface at the atomic level. These parameters have to be taken into account when predicting the reactivity of a given material. In general, the higher the surface area per unit mass, the higher the surface reactivity; this is the result of the higher proportion of surface atoms and their associated 'unsatisfied valences'. It follows that data relating to surface reactions with

small molecules on different fibre types should be compared per unit of exposed surface.

When comparing the effects of different kinds of fibres, the question arises of how to express exposure *in vivo* and doses *in vitro*. Currently, mass is often used, but alternatives include the number of fibres, the number of sized fibres and the unit surface area of the fibres. The appropriate selection depends on the biological process under investigation. If small molecules are thought to be acting as mediators of the biochemical reactions taking place, then exposed surface area would be the most appropriate. If cellular events or internalization of fibres are involved, a sized number of fibres (a size distribution) should be considered. The mass of the fibres, however, is generally the most inappropriate unit to use, particularly when comparing fibres of different specific weights or different surface-area-to-mass ratios. By means of illustration, differences found between crocidolite and erionite in some toxicity tests may be ascribed partially to the much larger exposed surface of the erionite (due to the open cage structure of the zeolite). Similarly, man-made vitreous fibre (MMVF) 10 glass fibre has an exposed surface area one order of magnitude lower than most asbestos fibres.

The fibre size distribution can be reported per number of fibres or per mass, and it is not always clear which should be used. However, agglomerations of fibres may skew the results – a few large, heavy particles or agglomerates will represent a negligible number of big particles using the first unit, but a fairly conspicuous mass in the second (Fubini *et al.*, 1995a). It should be noted that sonication can either create or disrupt agglomerations, and can cause fibre rupture.

Crystallinity. Crystallinity is evaluated by means of X-ray diffraction of the material or by selected area electron diffraction (SAED) of a single fibre under the electron microscope beam.

Natural mineral fibres – asbestos, zeolites and other fibrous silicates – are mostly crystalline, with amorphousness restricted to the outermost atomic layer. Mechanical, thermal or chemical treatments may cause amorphousness at depth. A simple chelating agent, desferrioxamine, partly destroys the crystalline structure even of amphibole asbestos by selective extraction of iron and other ions (Hochella, 1993; Mollo *et al.*, 1994), leaving an amorphous layer similar to that reported some time ago for crocidolite incubated in human blood serum (Crawford, 1980).

Artificial fibres are mostly amorphous (e.g. glasswool, slagwool and rockwool) as they are manufactured by rapid solidification of melt. Some ceramic materials – alumina, zirconia, carbides – are crystalline. They are often used in form of whiskers, i.e. a monocrystal that is elongated along one crystalline axis, a form that may be particularly hazardous because of the sharp, needle-like structure of fibres of respirable size. An example of this effect is in the pathogenicity of silicon carbide (SiC) whiskers, which has been investigated *in vitro* (Birchall *et al.*, 1988; Johnson *et al.*, 1992) and *in vivo* (Lapin *et al.*, 1991). On a fibre-for-fibre basis, SiC whiskers were found to be more cytotoxic than crocidolite (Johnson *et al.*, 1992); in contrast, fibre-free isometric SiC particles have been reported to be inert following inhalation studies, in-vitro cell tests and long-term injection tests (Bruch *et al.*, 1993a,b).

Micromorphology. The shape of the fibre (straight or curled) and the surface geometry (sharp or smooth truncations, indented edges) can be examined by TEM or scanning electron microscopy (SEM). High-resolution electron microscopy (HRTEM) provides a detailed microtopographic image of the particle surface at the atomic level.

Fractures along cleavage planes, steps and edges are the sites where surface reactions will mostly occur (Fubini, 1993; Hochella, 1993). An indented surface reacts more readily and to a larger extent than a smooth one of the same chemical composition.

Chemical composition
Elemental analysis. Elemental analysis of the fibre content is performed on aqueous solutions, in which the fibre has been dissolved, either by spectroscopic techniques or by chemical reactions, typically the precipitation of oxides. The former, which is more precise, gives an elemental percentage only. The latter furnishes separate data for different oxidation states (e.g. for iron: as ferrous oxide, FeO, and ferric oxide, Fe_2O_3).

Analysis of the chemical composition of the surface of the fibre is carried out using a variety of techniques, such as electron dispersive analysis by X-rays (EDAX), secondary ion mass spectrometry (SIMS) and X-ray photoemission spectroscopy (XPS).

Constitutive elements or contaminants: the case of iron in asbestos. Certain metal ions (typically iron) have the potential for free radical generation, a process that may be associated with the carcinogenicity of iron-containing minerals (reviewed by Kamp et al., 1992; Fubini, 1993; Hardy & Aust, 1995a).

Natural mineral fibre samples are very ill-defined because of the impurities that occur in variable concentration and intimate association with fibrous and nonfibrous minerals (e.g. nemalite, tremolite, magnetite in various chrysotiles); from the viewpoint of iron bioavailability, some of the unusual asbestiform contaminants appear to be potentially dangerous (Astolfi et al., 1991).

We may classify fibres according to their iron content in the following three main categories:

- The iron is a constitutive component of the mineral fibre, as in crocidolite $[(Na^+)_2(Fe^{2+})_3(Fe^{3+})_2(Si,Al,Fe^{3+})_8O_{22}(OH)_2]$ or amosite $[(Fe^{2+}Mg)_7(Si_8O_{22})(OH)_2]$, which are both amphiboles. The metal is present, in a rather large proportion, in a quantity mostly determined by the chemical formula.
- The iron substitutes for another ion. For example, in chrysotile asbestos $[Mg_3Si_2O_5(OH)_4]$, Fe^{2+} and Mn^{2+} can replace for Mg^{2+}. The quantity of substituted iron will vary from one specimen to another; low concentrations may also be found. Chrysotile asbestos is often associated with other iron-containing fibrous minerals, such as nemalite $[Mg(OH)_2]$, a fibrous brucite in which iron replaces magnesium in a relatively large proportion. Erionite, as a zeolite, contains several metal ions; iron may occasionally be present.
- The iron is present as an undesired impurity (in man-made fibres). This is often the case with rockwool and slagwool obtained from melt minerals; in traces, iron has also been found in glasswool. In these cases, the metal is not well dispersed over the fibre, but accumulates in some parts of it during manufacturing.

It has to be pointed out that not all iron contained in a fibre is potentially toxic (Fubini et al., 1995b), and that with some minerals, such as Canadian chrysotile, contaminants (e.g. tremolite or nemalite) may also play some role in toxicity.

Chemical composition at the surface and bioavailability. The chemical composition at the surface of a fibre is often different from that in its bulk (Fubini, 1993). Metal ions at the surface are in a higher oxidation state than in the bulk because of the spontaneous oxidation of surface ions (which are in contact with air). The rate of this process depends upon the location of the ion at the surface. Both of the oxidation states of iron, Fe(II) (ferrous) and Fe(III) (ferric), occur in crocidolite, while only Fe(II) is present in amosite. However, contrary to what is expected, the surface of amosite is poorer in Fe(II) than that of crocidolite (Shen et al., 1995), because the surface iron is more readily oxidized.

Newly created surfaces (e.g. from the grinding or simple breaking of the fibre) will expose Fe(II) which will be slowly oxidized in the atmosphere (Fubini et al., 1991). This may also account for the higher toxicity of freshly ground minerals. In contrast, heating accelerates oxidation so that artificial fibres obtained from melt will be fully oxidized at the surface.

Surface composition can be analysed by XPS, SEM and laser-assisted microprobe mass spectrometry analysis (LAMMA). However, these techniques provide analysis of the very outermost surface layers of a fibre, and this information can be rather imprecise, particularly in fibres with an extensive micromorphology. An indirect method for studying the chemical composition of the full exposed surface topography is through the adsorption of probe molecules.

The identification of one component implied in fibre toxicity, however, is not sufficient to predict the extent of damage and toxicity of other fibres of a similar composition. The mechanism of toxicity is initiated by complex interactions of solid and biological matter – the toxic moiety has to be bioavailable in order to elicit a biological response. Bioavailability depends on a number of factors related partly to the solid (e.g. redox state, crystallinity, micromorphology of the particle, location of the ions in the external layers) and partly to the biological compartments in which the fibre will be located (e.g. extracellular or intracellular medium, phospholipid lining layer, phagolysosome, etc.).

Effect of milling
Milling, a widely used preparatory method in experimental investigations of fibres, affects both

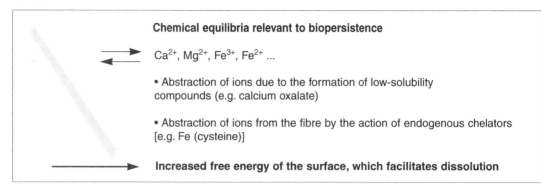

Figure 3. A fibre in an aqueous solution designed to mimic body fluids. An equilibrium is established in a closed system between the solid and dissolved ions that may be displaced by other solutes reacting with solubilized components of the fibre, usually metal ions. Erosion of the fibre at some surface points causes an increase in the free energy of the surface, which facilitates reactivity and dissolution.

the form and surface composition of fibres, and these changes should be borne in mind when interpreting data. Crystalline fibres (typically asbestos) retain a fibrous form after milling whereas amorphous ones (e.g. glass fibres) are progressively transformed into isometric particles. In both cases, newly cleaved surfaces are brought in contact with air and react, sometimes very slowly, with atmospheric components. Fe(II) is progressively oxidized to Fe(III), with kinetics strictly dependant on the crystal structure (Fubini et al., 1991). Asbestos fibres are usually activated by grinding, because Fe(II) is exposed. A remarkable increase in ROS release and in the formation of 8-hydroxydeoxyguanosine (8-OHdG) from deoxyguanosine was reported for crocidolite, amosite, anthophyllite and wollastonite following milling (Nejjari et al., 1993). In contrast, the milling of crocidolite asbestos did not affect iron-mediated DNA damage and lipid peroxidation (Turver & Brown, 1987). Pure silica is also activated by milling; some other fibres, however, lose their potential for free radical release, probably because milling favours full oxidation of iron and stabilization of Fe(III) in the solid matrix. Several studies using in-vitro cell systems have detected a decrease of fibre activity following milling (Ellouk & Jaurand, 1994).

A further important point is that some reactions only occur under mechanical stress (mechanochemistry), the product of which may be related to the following: (i) the chemical composition of the grinding chamber; (ii) the atmosphere in which milling took place (wet, dry, aerobic, anaerobic); (iii) the crystal structure of the fibre (Volante et al. 1994). Milled samples of fibres should be regarded as not necessarily identical in surface composition to unmilled samples.

When comparing data obtained with milled and unmilled fibres, the weight as well as the number of particles should be indicated.

Durability in vitro

Solubility tests. Mineral and man-made fibres are not readily soluble. As with any solid material, however, an equilibrium is established in water between the solid and its dissolved components (Fig. 3). This process is governed by the structure and chemical composition of the surface and is influenced markedly by pH and by the presence of other solutes. To attempt to mimic the situation *in vivo*, experiments can be performed with simulated body fluids (Chouikhi, 1995), the careful composition of which is very important. Because of the establishment of several fibre-reingestion cycles by alveolar macrophages (Fig. 2), extracellular, cytoplasmic and phagolysosomal fluids all have to be considered.

Solubility tests are made either with a static or dynamic assay, the latter obtained with fluxing solutions. A complete characterization of the process is done by measuring the following: weight loss; concentration of each component in the solution; and modifications in micromorphology (by SEM and TEM) and in surface composition (by XPS) (Baillif & Touray, 1994)

The situation *in vivo* is best considered to be intermediate between an open and a closed system. It is not clear whether either dynamic or static solubility assays reflect this situation more closely.

Selective leaching and ion exchange. Beside the mere solubility of the fibre, two other processes take place at the solid–liquid interface in biological fluids, namely, selective leaching and, with zeolites or layered silicates, ion exchange. Selective leaching is a process whereby one particular component of the solid, typically metal cations, are either seized by a chelator and brought into solution or react with an anion forming a compound of lower solubility which then precipitates. In both cases, the original fibre is progressively dissolved (Fig. 3). Very often such processes are influenced by reducing/oxidizing agents whereby the ion, by changing its size following redox reactions, is more easily extruded from the crystalline matrix (Hering & Stumm, 1990). More than one ion species can be removed by the same chelator; desferrioxamine is even more effective in removing magnesium than iron from crocidolite (Hochella, 1993).

Zeolites are typically associated with metal ions, which compensate for the negative surface charge of zeolites. These ions are 'exchangeable' – they may be displaced by other ions present in solution in higher concentration. An erionite fibre may thus acquire different kinds of metal ions, depending on the solutions with which it has been in contact.

Assays for the surface reactivity of fibres
Adsorption
Endogenous material and fibre coating. By the time an inhaled fibre reaches an alveolar region, contact with body fluids will have modified the external surface (via the physical adsorption of macromolecules at the very least). Surfactant treatments have been performed that attempt to model the initial events occurring when a particle deposits on the pulmonary alveolar hypophase. The effect of dipalmitoyl phosphatidylcholine (DPPC) surfactant on the genotoxic activity of chrysotile was investigated with the micronucleus assay on V79 cells (Lu *et al.*, 1994). The DPPC surfactant treatment was adequate to suppress the membranolytic activity, which is consistent with previous results. The treatment diminished but did not eliminate the activity of short fibre samples in the induction of micronuclei or multinucleated cells and did not significantly change the activity of fibres of intermediate length. The different behaviour of these short and intermediate fibres might be due to a difference in the stability of the coating of the two lengths of fibres; no difference was found between untreated short and intermediate-length fibres.

Differences in fibre length somehow involve a different extent of macromolecule binding at the surface. A dramatic enhancement of release of superoxide anions from rat alveolar macrophages was seen with long, but not with short, opsonized amosite fibres (Hill *et al.*, 1995). Besides the effect of incomplete phagocytosis, the increased biological activity of the long fibre sample can be explained by the increased binding of the opsonine to the fibre surface.

Exogenous substances: the fibre as carrier of chemical carcinogens. Prior to inhalation, any fibre may have adsorbed molecules, some toxic to humans, from the environment. Adsorption of polycyclic aromatic hydrocarbons (PAH) has been regarded as a possible explanation for the synergistic effects of asbestos exposure and tobacco smoking on the risk of lung cancer (Lakowicz & Hylden, 1978; Fournier & Pezerat, 1986; Gerde & Scholander, 1988; Fournier *et al.*, 1989). A much higher adsorption capacity for PAH was found for asbestos, in comparison with MMVF. Adsorption was decreased if it was carried out in the presence of water vapour, indicating competition between H_2O and PAH for adsorption sites. Several other explanations have been proposed for the synergistic effect of asbestos and cigarette smoke (Saracci, 1977; Mossman *et al.*, 1983; McFadden *et al.*, 1986; Hobson *et al.*, 1990). There is experimental evidence that asbestos and cigarette smoke synergistically increase DNA damage by stimulating the formation of free radicals (Jackson *et al.*, 1987). In any case, a high concentration of PAH at the fibre surface may enhance any reaction taking place between asbestos and cigarette smoke.

Generation of free radicals
How fibres give rise to free radicals. There is considerable evidence from in-vivo and in-vitro tests supporting the hypotheses that free radicals and other ROS are an important mechanism by which asbestos and other mineral fibres mediate genetic damage (Goodglick & Kane, 1986; Shatos *et al.*, 1987; Mossman *et al.*, 1990; Guilianelli *et al.*, 1993). Nevertheless, the mechanism whereby the fibre promotes or catalyses the abnormal release of radicals is still under debate. In asbestos, a crucial

role is played by iron (reviewed by Kamp et al., 1992; Hardy & Aust, 1995a). In glasswool, slagwool and rockwool, reactivity similar to asbestos has been reported by some authors (Gulumian & Van Wyk, 1987; Leanderson et al., 1988; Pezerat et al., 1992). Recent investigations on the transforming potency of various iron-containing solids on Syrian hamster embryo (SHE) cells revealed the following: (i) the transforming potency is related to iron, as it is markedly decreased by desferrioxamine; and (ii) not all iron is active (Elias et al., 1995).

If iron and ROS are implied in asbestos toxicity, the question arises as to the role of iron at the molecular level and whether the origin of ROS is exogenous or endogenous, or both. In fibre–cell interactions there are two following possible sources of free radicals:

- the fibre *per se* triggers radical reactions in aqueous suspensions and thus when in contact with body fluids;
- during phagocytosis of the fibre, ROS (including the superoxide anion and hydrogen peroxide) which are produced by the cell in the attempt to destroy the foreign body are released into the medium (Vallyathan et al., 1992).

It has recently been reported that asbestos fibres up-regulate nitric oxide production by rat alveolar macrophages (Thomas et al., 1994). Hydroxyl radicals (OH$^\bullet$) and peroxynitrite may be generated from nitric oxide.

In both of the above mechanisms, ROS may overwhelm antioxidant defences and, consequently, radicals may reach target cells and initiate a pathogenic process. There is evidence for both mechanisms occurring with some fibres, which implies, on the one hand, ubiquitous radical release at the fibre surface and, on the other, complex reactions occurring during internalization between radicals of 'inorganic' and 'biochemical' origin. Both processes are related to the surface chemistry of the fibre and a relationship has been reported between free radical release during the phagocytosis of various fibres and the cytotoxicity of the fibres (Vallyathan et al., 1992; Vallyathan, 1994).

Various hypotheses have been proposed regarding (i) the chemical nature of the iron sites active at the surface of fibres and (ii) the modifications of these iron sites *in vivo* following radical release (reviewed in Fubini & Mollo, 1995; Fubini et al., 1995b). Bearing in mind that all fibre-related diseases (malignant mesothelioma, bronchogenic carcinoma and asbestosis) are long-term ones, the molecular mechanism proposed should either trigger a cascade of reactions or provide a persistent release of toxic molecules likely via a catalytic process.

Iron may be either reactive at the solid surface or brought into solution by endogenous chelators. It has been demonstrated that single-strand DNA breaks correlate with the presence of species of iron that can be mobilized from asbestos by low molecular weight chelators (reviewed in Hardy & Aust, 1995a). This would suggest that free radical release correlates with mobilized iron. Mobilized iron is certainly a good candidate for acute damage as it can reach the target cell more easily than surface iron. However, a prolonged or catalytic mechanism consistent with long-term pathogenicity probably requires both surface and mobilized iron. It has recently been reported that the extent of DNA damage is not related to the amount of iron released in solution from a variety of asbestos and artificial fibres (Gilmour et al., 1995). Damage was proposed to stem from some iron at the fibre surface.

The following questions arise at this point:

- Is iron implied in radical release from any type of fibre or only from asbestos and some artificial fibres (glasswool, slagwool, rockwool)?
- As only some iron at the fibre surface is active (Fubini et al., 1995b; Gilmour et al., 1995), what is the oxidation and coordination state of the iron at the active site?
- As not all radicals are generated via a catalytic mechanism (Fubini et al., 1995b), under what circumstances can the fibre surface act as a heterogeneous catalyst for radical production *in vivo*?
- How do chelators modify the potential for radical release of a fibre? Is it through iron depletion, strong adsorption of the chelator or variations in the redox potential of iron?
- Does iron-catalysed ROS stem from iron within the solid, from that at the solid–liquid interface or from that mobilized by endogenous chelators?
- Iron in artificial amorphous fibres is in a different coordination and redox state than iron in crystalline asbestos. Can this account for

the different carcinogenic potential between these two materials?

Detection of free radicals: spin trap technique, use of scavengers and chemical reactions. The high reactivity of free radicals makes their detection difficult. Various methods are employed for oxygen-derived free radicals in biological systems (reviewed by Mason et al., 1994). Direct detection by electron paramagnetic resonance (EPR) is not applicable to the few and short-lived radicals that originate in aqueous fibre suspensions. Most evidence has been provided either by the spin trapping technique (Weitzman & Graceffa, 1984; Gulumian & Van Wyk, 1987; Zalma et al., 1987a,b; Vallyathan et al., 1992; Fubini et al., 1995b) or by analysis of the products of subsequent chemical (Johnson & Maples, 1994) or biochemical (Leanderson et al., 1988, 1989; Berger et al., 1993) reactions.

Spin trapping is a technique in which a short-lived reactive free radical combines with a diamagnetic molecule (a 'spin trap') to form a more stable free radical (the 'radical adduct') which can then be detected by EPR. Accurate analysis of the spectral features enables identification of the original radical.

Spin trapping was employed in earlier work to provide evidence that asbestos catalyses the generation of OH$^\bullet$ from hydrogen peroxide (Weitzmann & Graceffa, 1984). The radical reactivity in this case was strictly related to the presence of hydrogen peroxide, thus confined, *in vivo*, to phagocytic cells. It was subsequently reported that even in the absence of hydrogen peroxide some radicals are produced by asbestos (Zalma et al., 1987b). Other iron minerals – from mines where an excess of lung cancer had been found – were similarly reactive. A correlation was hypothesized between the presence of bioavailable Fe(II) and carcinogenicity (Costa et al., 1989).

Often a secondary trap is employed, such as formate, which gives more stable adducts than the OH$^\bullet$, based on the simple reaction:

$$OH^\bullet + HCO_2^- \rightarrow H_2O + CO_2^{-\bullet} \qquad [1]$$

However, this is somewhat misleading because a direct reaction not mediated by OH$^\bullet$ may also take place between iron and formate (see below). Also, in some cases, more than a single radical may originate from the solid. A spectrum is then originated by the superposition of lines from two or more adducts formed by the radicals with the trap.

Mechanisms of release: the role of d-ions, particularly iron, interfacial phenomena and/or mobilization of ions. The following three possible mechanisms of free radical release from aqueous suspensions of mineral fibres (cell-free tests) have been envisaged so far (discussed in Fubini et al., 1995b):

Mechanism I – production of OH$^\bullet$ in the presence of hydrogen peroxide, following a Fenton type reaction:

$$M^{n+} + H_2O_2 \rightarrow M^{(n+1)} + OH^\bullet + OH^- \qquad [2]$$

First proposed for asbestos (Weitzman & Graceffa, 1984) and subsequently also for glass fibres (Gulumian & Van Wyk, 1987) and for various other minerals in the presence of reducing agents (Kennedy et al., 1989), it applies only to reactions occurring in biological compartments where hydrogen peroxide is present, typically the phagolysosome.

Mechanism II – production of OH$^\bullet$ from O_2 present in the solution, following the reaction sequence:

$$O_2 + e^- \rightarrow O_2^{-\bullet} \qquad [3]$$

$$O_2^{-\bullet} + e^- + H^+ \rightarrow H_2O_2 \qquad [4]$$

$$H_2O_2 + e^- \rightarrow OH^\bullet + OH^- \qquad [5]$$

This mechanism was proposed for a large variety of mineral and artificial fibres (Zalma et al., 1987b; Leanderson et al., 1988; Pezerat et al., 1989). It may occur in any biological compartment, as oxygen is ubiquitous, but it requires efficient electron donating sites at the surface of the particle.

Mechanism III – hydrogen abstraction from an organic molecule, HR:

$$X + HR \rightarrow XH + R^\bullet \qquad [6]$$

This mechanism does not necessarily involve active oxygen species, although it does not take place in oxygen-free solutions. It appears to be the most general mechanism of the three as it comprises all ROS capable of hydrogen abstraction (e.g. OH$^\bullet$) and potentially even other reactive species. It has been reported for a large variety of fibres (Nejjari et al., 1993), iron minerals (Costa et al.,

1989) and iron-free hard metal particles (Lison et al., 1995). It is usually performed with formate generating a carboxyl radical (see equation 1, above).

Fubini et al. (1995b) reported that mechanisms I and III may occur on the same particle, but involve different kinds of active surface sites. The two mechanisms operate with different target molecules – H_2O_2 and HCO_2^- – and cause oxidation (Fig. 4a) and reduction (Fig. 4b) of the active surface site.

The three mechanisms apply to aqueous suspensions of solid fibres and involve interfacial phenomena. Iron may be strictly bound to the surface or mobilized at the interface layer and brought into the solution by chelators. In the experimental conditions reported for the evaluation of free radical release (spin trapping), the only possible chelator is the phosphate ion from the buffer. Investigations on the activity of the filtrate (Pezerat et al., 1989) have revealed that the effect of iron in solution is minimal, or absent. Filtrates in the presence of other chelators revealed some activity, but this was lower than that of the corresponding solid suspensions (Levasseur-Acker et al., 1995; Lison et al., 1995).

With asbestos, mechanism III is non-catalytic (i.e. in aqueous suspensions sites are consumed during the reaction) whereas Fenton activity (mechanism I) is catalytic (Fubini et al., 1995b).

Fibres pretreated with the same chelators as those used by L.G. Lund and A.E. Aust for the selective removal of Fe^{3+} and Fe^{2+} (desferrioxamine and ferrozine) fully inhibited radical release following hydrogen abstraction, which suggested that both oxidation states are required for mechanism III (Fubini et al., 1995b). Higher oxidation states (Fe=O, ferryl and perferryl) have also been proposed (Pezerat et al., 1992; Nejjari et al., 1993).

To distinguish between mechanisms II and III, the test has to be performed in the absence of formate to provide evidence for the role of OH• in mechanism II. As formate is used as a secondary trap (equation 1), available data do not allow such a distinction. On the other hand, the generation of the carboxylate radical from formate occurs mostly in aerobic conditions. The reaction itself, however, does not require oxygen per se. Is it correct to assume carboxylate radical release as a measure of oxygen consumption and ROS formation?

Not all iron-containing minerals are toxic; for example, iron oxides are not toxic and do not release free radicals in solution (Pezerat et al., 1989; Fubini et al., 1995b). This is not just a matter of oxidation state – magnetite (Fe_3O_4), like crocidolite, contains both Fe(II) and Fe(III). Thus there is evidence that iron becomes active only as part of some solid structures (Chouchane et al., 1994; Mollo et al., 1994; Fubini et al., 1995b).

It is important to investigate whether mechanism III, which is non-catalytic in cell-free systems, becomes catalytic in vivo, by cyclic activation of the surface active sites. There is some evidence that reductants reactivate the fibres (B. Fubini & L. Mollo, unpublished results). Questions arise about the possible role of endogenous reducing agents.

The carcinogenic potential of iron-free fibres such as erionite and ceramic fibres has not been explained yet. Either a different biochemical mechanism should be hypothesized or an interaction in vivo between endogenous iron and some surface sites takes place.

Modification of surface properties and use of model solids. The exact nature of the iron at surface sites implied in the above reactions is still undefined. There is a general consensus on the following points:
• Fe(II) is implied;
• iron is present in more than one oxidation state, most likely in clusters with an intermediate redox state (e.g. Fe(II)–Fe(III));

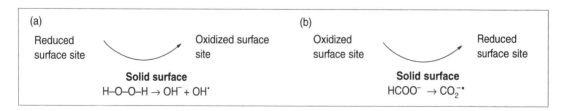

Figure 4. (a) Mechanism I and (b) mechanism III of free radical release from aqueous suspensions of mineral fibres in cell-free tests. Both mechanisms I and III may occur on the same fibre, but they involve different target molecules, H_2O_2 in mechanism I and HCO_2^- in mechanism III.

- ligands in solution (as well as the solid surface matrix acting as a ligand) are thought to modify the redox potential and modulate the potential for free radical release, which means that iron will behave differently in fibres of different crystallochemistry;
- only a fraction of the iron at the surface is active; and
- iron loads do not correlate with the potential for free radical release – an inverse correlation (i.e. maximum activity for well-dispersed surface iron) might even be envisaged, as in one case glass fibres were more active than crocidolite (Gulumian & Van Wyk, 1987).

Surface-modified fibres have been used to clarify these points. For example, crocidolite was enriched in either Fe(III) or Fe(II) in a manner that altered the surface chemistry of the fibres without changing their morphology (Gulumian et al., 1993a,b). Fe(II) was correlated with iron mobilized by ferrozine and with the extent of free radical release (spin trap technique); Fe(III) somehow 'detoxified' the fibre. In another study, an increase in DNA single-strand breaks was found with crocidolite incubated in tissue culture medium containing iron salts (Hardy & Aust, 1995b). The effect of iron acquisition by erionite on the induction of single-strand breaks in DNA was related to the redox state of iron. When Fe(II) was bound to erionite, single-strand breaks in DNA were induced in the absence of ascorbate, but when Fe(III) was bound, DNA single-strand breaks were observed only in the presence of ascorbate (Eborn & Aust, 1995).

An Fe(II)-exchanged zeolite was employed in the laboratory of L. Mollo and B. Fubini as a model solid for a systematic study on the relationship between iron on a solid surface and free radical release. It revealed that mechanism III (hydrogen abstraction) is related to the presence of Fe^{2+} at the surface whereas Fenton activity (mechanism I) is not (Fubini et al., 1995b). Full oxidation of the surface inhibits hydrogen abstraction, but treatment of the particle with glutathione may restore it (L. Mollo & B. Fubini, unpublished results).

Iron-loaded glass microbeads were employed in cellular tests to simulate the possible role of iron in glasswool and rockwool fibres. No iron-related toxicity was evident (Dong, 1994). Whether this indicates that iron in a glass matrix is not active, or that iron in this case had been fully oxidized and inactivated during manufacturing, has to be clarified. Still to be investigated, however, is whether trace iron – even at levels that cannot be detected by chemical analysis – may induce biological responses from artificial fibres similar to those triggered by iron in asbestos.

Radicals from non-iron-containing fibres. Not all harmful fibres release radicals nor contain iron. Most ceramic ones do not. SiC whiskers (Birchall et al., 1988) and fibrous silicon nitride (Si_3N_4) (Fisher et al., 1989) are toxic using in-vitro and in-vivo tests. A mechanism whereby free radicals originate from either the activity of iron or other components deposited (even in small traces) during manufacture may be invoked. Both SiC and Si_3N_4 exhibit a substantial number of silicon-derived surface radicals, whose abundance increases after grinding. SiC is active in hydrogen abstraction (Fubini et al., 1991); however, a relationship between surface and released radicals has not yet been proved. In hard metal particles, cobalt is an active generator of free radicals (Lison et al., 1995), and for this reason cobalt impurities should also be regarded as a possible source.

The extent of OH$^\bullet$ originating from tungsten oxide (WO_3) fibres, as measured by their reaction with deoxyguanosine, was close to that of crocidolite and higher than that of anthophyllite on a weight-for-weight basis (Leanderson & Sahle, 1995). Catalase substantially reduced radical formation, as did desferrioxamine, although to a lesser extent. This would suggest that radicals may originate both from WO_3 and some trace iron.

Biological targets of free radicals in cell-free systems
DNA damage. As reported above, ROS is postulated to mediate DNA damage induced by fibre suspensions. DNA damage caused by asbestos has been investigated in various ways, mostly by measuring single-/double-strand breaks or the oxidation of purine bases. Production of 8-OHdG was detected both in the presence (Kasai & Nishimura, 1988) and in the absence (Leanderson et al., 1988) of hydrogen peroxide. In spite of OH$^\bullet$ being a short-lived radical, direct hydroxylation of DNA by OH$^\bullet$ generated in the presence of asbestos was evident. A mineral-mediated oxidation yielding purine decomposition products was obtained by incubation

of minerals with nucleosides (Berger et al., 1993). A comparison of the ability of a fibre to generate $CO_2^{-\bullet}$ via hydrogen abstraction (mechanism III, see above) and the formation of 8-OHdG also yielded a very strong correlation (Nejjari et al., 1993) with several kinds of active rockwool and glasswool and non-active fibres (see *Comparison of different natural and man-made fibres*).

In the presence of organic peroxides and hydroperoxides, asbestos-mediated DNA damage was increased compared either to asbestos alone or to peroxide/hydroperoxide alone (Mahmod et al., 1993). Both ROS scavengers and desferrioxamine protected against DNA damage. As with hydrogen peroxide, double-strand breaks and enhanced DNA fidelity were observed, but sugar damage was only found with hydrogen peroxide (Mahmod et al., 1994). This different behaviour was ascribed to the differential reactivity of DNA with $OH^{\bullet}$ and alkoxy/aryloxy free radicals.

Crocidolite-dependent single-strand breaks in DNA showed a strong correlation with iron mobilized from these fibres (reviewed in Hardy & Aust, 1995a). Iron mobilized from asbestos by low molecular weight chelators (see below) will reach the target cell, especially DNA, more easily than that fixed at the surface. Hydrogen peroxide and ascorbate increased the induction of single-strand breaks, suggesting a redox radical mechanism involving iron in a low oxidation state.

It is not clear which of the three radical mechanisms outlined above correlates with DNA damage. There is evidence for a correlation with non-catalytic hydrogen abstraction (Nejjari et al., 1993), but this does not mean that other radicals might not also be active. Correlations found for the activity of a series of fibres in various assays do not necessarily mean that the same chemical functionalities are implied.

Although referred to as 'asbestos-catalysed DNA damage', there is no evidence for the damaging reaction really being a catalytic one. Repeated DNA damage assays on the same sample, in fact, have never been reported. A clear-cut distinction has to be made between catalytic, renewable and consumable sites. By renewable sites, we mean sites that react releasing DNA-damaging moieties but are recovered by other reactions (e.g. redox cycling, ion migration).

As direct evidence of radicals is obtained by the spin trapping technique, which is always performed in aqueous buffered solutions, it is not possible to state whether radicals stem from iron embedded in the outermost solid layers, from iron bound loosely at the solid–liquid interface or from iron just brought into solution by chelators (e.g. the phosphate buffer). Tests on the filtrate failed to detect appreciable amounts of radicals originating from solubilized iron (see above). Assays of DNA damage are more sensitive than spin trapping. Considering that the fibre has to be in close proximity to the target molecule to catalyse damaging reactions, target cells will be more easily damaged by mobilized iron. It can thus be hypothesized either that the few radicals that originate in solution, which escape detection by EPR, are responsible for DNA damage or that both surface and solubilized iron are active via an as yet unknown interplay between the two.

Lipid peroxidation. There is considerable evidence that asbestos causes lipid peroxidation in cell membrane lipids or in model systems such as fatty acid emulsions or liposomes; lipid peroxidation is often linked to iron-derived ROS, because ROS scavengers and desferrioxamine inhibit the reaction (reviewed by Kamp et al., 1992). Free radical generation via hydrogen abstraction (mechanism III, outlined above), however, is not correlated with lipid peroxidation (Fontecave et al., 1990). The two processes have common characteristics in that they both depend on iron ions available in the mineral and both require Fe^{2+}, but a comparative analysis of the results obtained on the same group of iron-containing minerals revealed different activities in the two tests. Crocidolite-induced lipid peroxidation was also unrelated to macrophage toxicity (Goodglick et al., 1989). It has been proposed that there are two types of ROS generated by iron – one, active in mechanism III and in lipid peroxidation; the other one, less electrophilic, active only in lipid peroxidation (Fournier et al., 1991, 1995).

It has been reported recently that fibre-induced lipid peroxidation depends both on the composition of the incubating solution (mainly the presence of chelators) and on the texture of the fibres involved. These factors may also explain why some fibres induce lipid peroxidation in cell free tests but not on epithelial cell membranes (Levasseur-Acker et al., 1995). Wollastonite, apparently iron-free, has also been found to be active in lipid peroxidation (Aslam et al., 1992).

As more than one type of ROS has been found, future investigations should focus on elucidating which mechanism is related to lipid peroxidation. There is evidence for Fenton activity being related to it, but is not clear whether hydrogen peroxide is required or not.

Mobilization and deposition of iron
Iron depletion. Removal of iron from asbestos by incubation in several iron chelators was widely investigated by A.E. Aust and co-workers (reviewed in Hardy & Aust, 1995a). In spite of the minimal water-solubility of the mineral fibre, chelators induce a substantial release of iron. Association of a reductant with a chelator enhances iron release. Iron was also mobilized from asbestos fibres by cells in culture in association with low molecular weight proteins and chelators (Chao et al., 1994).

Mobilization was found to be a function, not only of the chelator, but also of surface area, crystalline structure and iron content (Lund & Aust, 1990). Iron released from different iron-containing minerals, however, does not parallel the total iron content but depends on several factors, including the coordination of the ion and redox state (Mollo et al., 1994). The stability constant of various chelators does not correlate with the rate of mobilization of iron from crocidolite – size and geometry of the chelator also have to be considered (Hardy & Aust, 1995a). More iron was released from nemalite than from chrysotile or crocidolite, indicating that crystalline structure and not iron content governs mobilization kinetics (Fubini & Mollo, 1995). By comparing the results obtained on a set of fibres by a set of chelators, we have recently found that the potential for iron abstraction of a given chelator depends upon the solid involved, e.g. the most effective chelator for iron in crocidolite is not necessarily the most effective one for iron in amosite (B. Fubini, unpublished).

Prolonged (Chao et al., 1994) or successive (Mollo et al., 1994) incubations reveal chelator-induced ion mobility within the solid. A true equilibrium state between the solid and a chelating solution is virtually never attained. The crystalline structure, as detected by X-ray diffraction, was unaffected by chelation, but modifications did appear in micromorphology (Mollo et al., 1994). It is not easy to predict how much iron (or other ions) may be mobilized from a solid by a chelator. A study on crystal–chelator interaction and on the relationship between crystal structure and iron release kinetics should be undertaken.

Iron deposition and asbestos bodies. Iron-depleted fibres may take up iron from solution. Such iron coating has been reported above (see *Generation of free radicals*). A peculiar case of deposition is the formation *in vivo* of ferruginous bodies, which are fibres coated by endogenous iron and whose biological role is unknown. These bodies represent a form of biomineralization originating from the lung defence, presumably in an attempt to isolate the fibre from the lung surface. Ferruginous bodies are considered to be a direct marker of fibre exposure. They contain considerable amounts of iron which may potentially become dangerous. Iron associated with amosite was found to be responsible for the formation of single-strand breaks in DNA in the presence of ascorbate and low molecular weight chelators to a larger extent than uncoated fibres of similar size (Lund et al., 1994).

Much of the data points to the complex cycling of iron at the fibre surface *in vivo*; iron is depleted by endogenous chelators and then re-deposition as ferritin and/or hemosiderin in a coating on the fibre (Fig. 5). If iron-derived ROS are related to carcinogenicity, this cycling will modulate the carcinogenic potential of the fibre.

Comparison of different natural and man-made fibres
Data on various fibres
A comparative study of the surface reactivity of most types of fibres in use has not yet been performed. Because of different protocols and origin of samples, comparison between experimental data from different laboratories is difficult. Tables 1 and 2 summarize the data on free radical release, DNA damage and lipid peroxidation that is available for the mineral and man-made fibres mentioned in this paper and the Consensus Report in this volume.

There is evidence for a correlation between the three processes examined and the presence of iron in mineral fibres, but, as pointed out in the text, the reaction mechanisms still need clarification and are not necessarily related to the same surface functionality.

Of the two pieces of evidence for DNA damage, single-strand breaks (investigated in the laboratory of A.E. Aust) and 8-OHdG (investigated in the laboratory of P. Leanderson), the former correlates with removable iron and is not enhanced by grinding, the latter is enhanced or could even be originated by grinding, as is the hydrogen abstraction mechanism III. A model biomacromolecule, supercoiled fX 174RFI plasmid DNA, has also recently been employed to detect free radical damage (Gilmour et al., 1995).

Basic research with simple model solids is required for the assignment of a given ROS to the effects on lipids and DNA. There are also no experimental data stating which of the above processes is catalytic in cell-free assays or may become catalytic *in vivo*; further research is needed using artificial fibres.

Iron content, free radical release and carcinogenic potential

The relationship between iron, free radical release and biological damage indicating a potential for carcinogenicity has been established at least with the most common asbestos fibres, namely crocidolite, amosite and chrysotile in cell-free (Leanderson et al., 1988; Fontecave et al., 1990; Kamp et al., 1992; Gilmour et al., 1995; Hardy & Aust, 1995a), cellular (Shatos et al., 1987; Vallyathan, 1994) and in-vivo tests (Mossman et al., 1990; Moyer et al., 1994). The carcinogenic potential of iron in these solids stems from its redox state and location within the solid. Free radical release, as measured in cell-free assays, is unrelated to iron content. This does not mean that, when it comes to carcinogenic potential, iron content might not play a role in providing a long-lasting and/or a renewable source of free radicals. This would explain why some fibres (e.g. glass fibres and crocidolite) exhibit a similar extent of free radical release in spite of their different carcinogenicity and iron content.

Some fibres (e.g. carbides) release radicals without appreciable amounts of iron; others, such as ceramic fibres, do not release any radicals but are carcinogenic. Only following iron loading does erionite, the carcinogenicity of which is well established, become active in free radical release.

Further research is needed to establish: (i) evidence of radical sources other than iron; (ii) the extent and biological impact of radicals released by small traces of iron in a fibre; and (iii) the presence of surface structures on non-radical-releasing carcinogenic fibres apt to accommodate and activate endogenous iron.

Limitations of physico-chemical assays

The major limitation in the use of physico-chemical assays resides in the fact that they have to be performed very rapidly and with concentrations much higher than those encountered *in vivo*. Once there is evidence of a given reaction occurring in an aqueous suspension of fibres in a cell-free assay, the following assumptions for extrapolating the result obtained to in-vivo systems have to be made:

• the same reaction occurs at much lower concentrations;
• the chemical model is consistent with long-term pathogenicity, i.e. the reaction is either catalytic or there is cycling of reactants *in vivo* that provides a long-lasting reactivity;

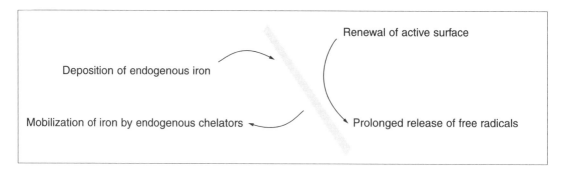

Figure 5. Iron cycling *in vivo* and surface reactions consistent with the prolonged release of free radicals.

Table 1. Mineral fibres: surface reactivity from cell-free tests

Fibre type	Free radical release mechanism[a]	DNA damage[b]	Lipid peroxidation	Reference
Amphibole asbestos				
Crocidolite	I		+	Kamp et al., 1992
	III[c]	8-OH-dG[c]		Nejjari et al., 1993
		SSB		Lund & Aust, 1992
		Supercoiled depletion		Gilmour et al., 1995
			+	Levasseur-Acker et al., 1995
Amosite	I			Kamp et al., 1992
	III[c]	8-OH-dG[c]		Nejjari et al., 1993
		SSB		Hardy & Aust, 1995a
		Supercoiled depletion		Gilmour et al., 1995
			+	Weitzman & Weitberg, 1985
Anthophyllite	I			Maples & Johnson, 1992
	III[d]	8-OH-dG[c]		Nejjari et al., 1993
Tremolite	I			Maples & Johnson, 1992
		SSB		Lund & Aust, 1992
Serpentine asbestos				
Chrysotile A	III[c]			Pezerat et al., 1989
		8-OHdG		Berger et al., 1993
			+	Levasseur-Acker et al., 1995
Chrysotile B	I		+	Kamp et al., 1992
	III[c]			Pezerat et al., 1989
		SSB		Lund & Aust, 1992
			+	Levasseur-Acker et al., 1995
Asbestiform fibres				
Wollastonite	III[c]	8-OHdG[c]		Nejjari et al., 1993
			+	Aslam et al., 1992
Nemalite	I, III[c]			Pezerat et al., 1989
		8-OHdG		Berger et al., 1993
			+	Fontecave et al., 1990
				Levasseur-Acker et al., 1995
Zeolites				
Erionite	I			Maples & Johnson, 1992
		[SSB][e]		Hardy & Aust, 1995a

Data are not available for the following mineral fibres: actinolite (amphibole); talc (asbestiform); mordenite (zeolite); attapulgite, sepiolite (clays).
[a]I, mechanism I – Fenton activity, hydroxyl radical (OH$^{\bullet}$) formation; III, mechanism III – hydrogen abstraction from formate.
[b]8-OHdG, 8-hydroxydeoxyguanosine – DNA base hydroxylation; SSB, single-strand breaks in DNA.
[c]Reactivity possibly originated or enhanced by milling of fibre sample.
[d]Inactive or poorly reactive.
[e]Only if iron loaded.

Fibre type	Free radical release mechanism[a]	DNA damage[b]	Lipid peroxidation	Reference
Vitreous (MMVF)				
Glasswool	I			Maples & Johnson, 1992
	III			Pezerat et al., 1992
Glass fibres	I			Gulumian & Van Wyk, 1987
				Maples & Johnson, 1992
		Supercoiled depletion		Gilmour et al., 1995
Rockwool	II[c]	8-OHdG[c]		Nejjari et al., 1993
		Supercoiled depletion		Gilmour et al., 1995
Slagwool	III			Pezerat et al., 1992
Ceramic	III[d]	+[d]		Nejjari et al., 1993
		Supercoiled depletion		Gilmour et al., 1995
Crystalline				
Calcium silicate	III[d]	+[d]		Nejjari et al., 1993
Potassium titanate	III[d]		+[d]	Nejjari et al., 1993
Silicon carbide	I[d]			Johnson & Maples, 1994
	III			Fubini et al., 1991
Tungsten oxide		8-OHdG		Leanderson & Sahle, 1995

Data are not available for the following man-made fibres: alumina, graphite, sodium-aluminium carbonate, zeolites (crystalline); para-aramid, cellulose (organic).
[a]I, mechanism I – Fenton activity, hydroxyl radical (OH•) formation; III, mechanism III – hydrogen abstraction from formate.
[b]8-OHdG, 8-hydroxydeoxyguanosine – DNA base hydroxylation.
[c]Reactivity possibly originated or enhanced by milling of fibre sample.
[d]Inactive or poorly reactive.

- there are no components in the living system that interfere and inhibit such a reaction.

This last point is particularly relevant, considering, on the one hand, all of the adsorption equilibria likely to be taking place in the cytoplasm and extracellular fluids and, on the other hand, all the possible quenchers of free radicals (e.g. albumin) beside the specific enzymes. Lipid peroxidation occurring in a cell-free test and not with cells (Levasseur-Acker et al., 1995) is a typical example.

With free radical reactions, questions arise as to how radicals reach their target molecules in vivo before radical decay.

Biopersistence is the result of many contrasting events such as the following: adsorption/desorption; leaching/deposition of ions; and periodic variations of pH during successive phagocytosis. It is vital to know the effect of each of these factors separately, but prediction of what will happen in vivo requires models capable of simulating all of these processes simultaneously.

There is nevertheless a need for much more physico-chemical data than are so far available, for the following two reasons:

- Cell-free tests validated against what is known about fibre carcinogenicity are the quickest and cheapest way to prescreen a large number of new fibrous materials. There is a consensus on the deleterious role of free radicals and on the related iron toxicity, which leads to suspicion of any fibre that originates radicals or contains 'active iron'.

• Knowledge of surface reactivity allows the preparation of model fibres for in-vivo and in-vitro tests with controlled surface properties. These would be the most direct route to the chemical identification of the active surface site responsible for carcinogenicity, the knowledge of which would allow the design of tailored, non-carcinogenic fibrous materials.

References

Adamson, I.Y.R., Bakowska, J. & Bowden, D.H. (1993) Mesothelial cell proliferation after instillation of long or short asbestos fibers in mouse lung. *Am. J. Pathol.*, 142, 1209–1216

Aslam, M., Ashquin, M. & Rahman, Q. (1992) In vitro cytotoxic effects of wollastonites on rat hepatocytes: II. Lipid peroxidation and glutathione depletion. *Bull. Environ. Contam. Toxicol.*, 49, 547–554

Astolfi, A., Belluso, E., Ferraris, G., Fubini, B., Giamello, E. & Volante, M. (1991) Asbestiform minerals associated with chrysotile from western Alps (Piedmont-Italy): chemical characteristics and possible related toxicity. In: Brown, R.C., Hoskins, J.A. & Johnson, N.F., eds, *Mechanisms in Fibre Carcinogenesis*, New York, Plenum Press, pp. 269–283

Baillif, P. & Touray, J.C. (1994) Chemical behaviour of aluminum and phosphorous during dissolution of glass fibers in physiological saline solutions. *Environ. Health Perspectives*, 102 (Suppl. 5), 77–81

Berger, M., de Hazen, M., Nejjari, A., Fournier, J., Guignard, J., Pezerat, H. & Cadet, J. (1993) Radical oxidation reactions of the purine moiety of 2′-deoxyribonucleosides and DNA by iron-containing minerals. *Carcinogenesis*, 14, 41–46

Birchall, J.D., Stanley, D.R., Mockford, M.J., Pigott, G.H. & Pinto, P.J. (1988) Toxicity of silicon carbide whiskers. *J. Mat. Sci. Lett.*, 7, 350–352

Bruch, J., Rehn, B., Song, W., Gono, E. & Malkusch, W. (1993a) Toxicological investigations on silicon carbide. 1. Inhalation studies. *Br. J. Ind. Med.*, 50, 797–806

Bruch, J., Rehn, B., Song, W., Gono, E. & Malkusch, W. (1993b) Toxicological investigations on silicon carbide. 2. In vitro cell tests and long term injection tests. *Br. J. Ind. Med.*, 50, 807–813

Chao, C., Lund, L.G., Zinn, K.R. & Aust, A.E. (1994) Iron mobilization from crocidolite asbestos by human lung carcinoma cells. *Arch. Biochem. Biophys.*, 314, 384–391

Churg, A. (1994) The role of active oxygen species in uptake of mineral particles by tracheobronchial epithelial cells. In: Davis, J.M.G. & Jaurand, M.-C., eds, *Cellular and Molecular Effects of Mineral and Synthetic Dusts and Fibres* (NATO ASI Series Vol. H 85), Berlin, Springer-Verlag, pp. 1–8

Costa, D., Guignard, J., Zalma, R. & Pezerat, H. (1989) Production of free radicals arising from the surface activity of mineral and oxygen. I. Iron mine ores. *Toxicol. Ind. Health*, 5, 1061–1078

Chouchane, S., Guignard, J., & Pezerat, H. (1994) Appearance of very electrophilic species generated by some iron oxides: effect of iron chelators and reducing agents. In: Davis, J.M.G. & Jaurand, M.-C., eds, *Cellular and Molecular Effects of Mineral and Synthetic Dusts and Fibres* (NATO ASI Series Vol. H 85), Berlin, Springer-Verlag, pp. 397–402

Chouikhi, B. (1995) *Etude de la Cinétique de Dissolution des Fibres de Verre Potentiellement Cancerigènes en Milieux Biologiques Simulés*, Thèse, Université d'Orleans

Crawford, D. (1980) Electron microscopy applied to studies of the biological significance of defects in crocidolite asbestos. *J. Microscopy*, 120, 181–192

Dong, H.Y. (1994) *Role des Caracteristiques Physiques et Chimiques des Fibres Minerales dans leur Cytotoxicité et Genotoxicité sur Cellules Mesotheliales Pleurales de Rat*, Thèse, Université Paris–Val de Marne

Dunningan, J. (1984) Biological effect of fibers: Stanton's hypothesis revisited. *Environ. Health Perspectives*, 57, 333–337

Eborn, S.K. & Aust, A.E. (1995) Effects of iron acquisition on induction of DNA single-strand breaks by erionite, a carcinogenic mineral fiber. *Arch. Biochem. Biophys.*, 316, 507–514

Elias, Z. Poirot, O., Schneider, O., Marande, R.A., Danière, M.C., Terzetti, F., Pezerat, H., Fournier, J. & Zalma, R. (1995) Cytotoxic and transforming effects of some iron-containing minerals in Syrian hamster embryo cells. *Cancer Detect. Prev.*, 19, 405–415

Ellouk, S.A. & Jaurand, M.-C. (1994) Revew of animal/*in vitro* data on biological effects of man-made fibres. *Environ. Health Perspectives.*, 102 (Suppl. 2), 47–54

Fisher, G.L., McNeill, K.L., Singer, A.W. & Smith, J.T. (1989). A comparison of the *in vitro* and *in vivo* toxicity of fibrous and non-fibrous silicon nitride. In: Mossman, B.T. & Begin, R.O., eds, *Effects of Mineral Dusts on Cells* (NATO ASI Series H, Vol. 30), Berlin, Springer-Verlag, pp. 27–36

Fontecave, M., Jauen, M., Mansuy, D., Costa, D., Zalma, R. & Pezerat, H. (1990) Microsomal lipid peroxidation and oxyradical formation are induced by insoluble iron-containing minerals. *Biochem. Biophys. Res. Comm.*, 173, 912–918

Fournier, J. & Pezerat, H. (1986) Studies on surface properties of asbestos. III. Interaction between asbestos and polynuclear aromatic hydrocarbons. *Environ. Res.*, 41, 276–295

Fournier, J., Fubini, B., Bolis, V. & Pezerat, H. (1989) Thermodynamic aspects in the adsorption of polynuclear aromatic hydrocarbons on chrysotile and silica – possible relation to synergistic effects in lung toxicity. *Can. J. Chem.*, 67, 289–296

Fournier, J., Guignard, J., Nejjari, A., Zalma, R. & Pezerat, H. (1991) The role of iron in the redox surface activity of fibres. Relation to carcinogenicity. In: Brown, R.C., Hoskins, J.A. & Johnson, N.F., eds, *Mechanisms in Fibre Carcinogenesis*, New York, Plenum Press, pp. 407–414

Fournier, J., Copin, E., Chouchane, S., Dzwigaj S. & Guignard J. (1995) Peroxidation de l'acide lipidique en présence de composés inorganiques. Relation avec les mécanismes de stress oxydant. *C. R. Seances Soc. Biol. Fil.*, 189, 429–442

Fubini, B. (1993) The possible role of surface chemistry in the toxicity of inhaled fibers. In: Warheit, D.B., ed., *Fiber Toxicology*, San Diego, Academic Press, pp. 229–257

Fubini, B. & Mollo, L. (1995) Role of iron in the reactivity of mineral fibres. *Toxicol. Lett.*, 82–83, 951–960

Fubini, B., Bolis, V., Giamello, E., & Volante, M. (1991) Chemical functionalities at the broken fibre surface relatable to free radicals production. In: Brown, R.C., Hoskins, J.A. & Johnson, N.F., eds, *Mechanisms in Fibre Carcinogenesis*, New York, Plenum Press, pp. 415–432

Fubini, B., Bolis, V., Cavenago, A. & Volante, M. (1995a) Physico-chemical properties of crystalline silica dusts and their possible implication in various biological responses. *Scand. J. Work Environ. Health*, 21 (Suppl. 2), 9–14

Fubini, B., Mollo, L. & Giamello, E. (1995b) Free radical generation at the solid/liquid interface in iron containing minerals. *Free Rad. Res.*, 23, 593–614

Gerde, P. & Scholander, P. (1988) Adsorption of benzo-(a)pyrene onto asbestos and man-made mineral fibres in an aqueous solution and in a biological model solution. *Br. J. Ind. Med.*, 45, 682–688

Gilmour, P.S., Beswick, P.H., Brown, D.M. & Donaldson, K. (1995) Detection of surface free radical activity of respirable industrial fibres using supercoiled fX 174RFI plasmid DNA. *Carcinogenesis*, 16, 2973–2979

Guilianelli, C., Baeza-Squiban, A., Boisvieux-Ulrich E., Houcine, O., Zalma, R., Guennou, C., Pezerat, H. & Marano, F. (1993) Effect of mineral particles containing iron on primary cultures of rabbit tracheal epithelial cells: possible implication of oxidative stress. *Environ. Health Perspectives*, 101, 436–442

Goodglick, L.A. & Kane, A.B. (1986) Role of reactive oxygen metabolites in crocidolite asbestos toxicity to mouse macrophages. *Cancer Res.*, 46, 5558–5566

Goodglick, L.A. & Kane, A.B. (1990) Cytotoxicity of long and short crocidolite asbestos fibers *in vitro* and *in vivo*. *Cancer Res.*, 50, 5153–5163

Goodglick, L.A., Pietras, L.A. & Kane, A.B. (1989) Evaluation of the casual relationship between crocidolite asbestos-induced lipid peroxiation and toxicity to macrophages. *Am. Rev. Respir. Dis.*, 139, 1265–1273

Gulumian, M. & Van Wyk, J. A. (1987) Hydroxyl radical production in the presence of fibres by a Fenton-type reaction. *Chem. Biol. Interactions*, 62, 89–97

Gulumian, M., Bhoolia, D.J., Du Toit, R.S.J., Rendall, R.E.G., Pollak, H., Van Wyk, J.A. & Rhempula, M. (1993a) Activation of UICC crocidolite: the effect of conversion of some ferric ions to ferrous ions. *Environ. Res.*, 60, 193–206

Gulumian, M., Van Wyk, J.A., Hearne, G.R., Kolk, B. & Pollak, H. (1993b) ESR and Mössbauer studies on detoxified crocidolite: mechanism of reduced toxicity. *J. Inorg. Biochem.*, 50, 133–143

Hardy, J.A. & Aust, A.E. (1995a) Iron in asbestos chemistry and carcinogenicity. *Chem. Rev.*, 95, 97–118

Hardy, J.A. & Aust, A.E. (1995b) The effect of iron binding on the ability of crocidolite asbestos to catalyze DNA single-strand breaks. *Carcinogenesis*, 16, 319–325

Hemenway, D.H., Absher, M., Fubini, B. & Bolis, V. (1993) What is the relationship between hemolytic potential and fibrogenicity of mineral dusts? *Arch. Environ. Health*, 48, 343–347

Hering, J.G. & Stumm, W. (1990) Oxidative and reductive dissolution of minerals. In: Hochella, M.F., Jr & White, A.F., eds, *Reviews in Mineralogy, Mineral-water Interface Geochemistry*, Vol. 23, Chelsea, MI, Bookcrafters, pp. 427–465

Hill, I.M., Beswick, P.H. & Donaldson, K. (1995) Differential release of superoxide anions by macrophages treated with long and short fibre amosite asbestos is a consequence of differential affinity for opsonin. *Occup. Environ. Med.*, 52, 92–96

Hobson, J., Wright, J.L. & Churg, A. (1990) Active oxygen species mediate asbestos fibre uptake by tracheal epithelial cells. *FASEB J.*, 4, 3134–3139

Hochella, M.F. (1993) In: Guthrie, G.D. & Mossman, B.T., eds, *Reviews in Mineralogy*, Vol. 28, Chelsea, MI, Bookcrafters, pp. 275–308

Jackson, J.H., Schaufstatter, I.U., Hyslop, P.A., Vosbeck, K., Sauerheber, R., Weitzman, S.A. & Cochrane, C.G. (1987) Hydroxyl radical mediates the synergistic DNA damaging effects of asbestos and cigarette smoke. *J. Clin. Invest.*, 80, 1090–1095

Jaurand, M.-C. (1989) Particulate-state carcinogenesis: a survey of recent studies on the mechanism of action of fibres. In: Bignon, J., Peto, J. & Saracci, R., eds, *Non-Occupational Exposure to Mineral Fibres* (IARC Scientific Publications No. 90), Lyon, IARC, pp. 54–73

Johnson, N.F. & Maples, K.R. (1994) Fiber-induced hydroxyl radical formation and DNA damage. In: Davis, J.M.G. & Jaurand, M.-C., eds, *Cellular and Molecular Effects of Mineral and Synthetic Dusts and Fibres* (NATO ASI Series Vol. H 85), Berlin, Springer-Verlag, pp. 23–37

Johnson, N.F., Hoover, M.D., Thomassen, D.G., Cheng, Y.S., Dalley, A. & Brooks, A.L. (1992) In vitro activity of silicon carbide whiskers in comparison to other industrial fibers using four cell colture systems. *Am. J. Ind. Med.*, 21, 807–823

Kamp, D.W., Graceffa, P., Pryor, W.A. & Weitzman, S.A. (1992) The role of free radicals in asbestos-induced diseases. *Free Rad. Biol. Med.*, 12, 293–315

Kasai, H. & Nishimura, S. (1984) DNA damage induced by asbestos in the presence of hydrogen peroxide. *Gann Monogr. Cancer Res.*, 75, 841–844

Kennedy, H., Hopkins, C., Baser, M., Tolley, E. & Hoidal, J.R. (1989) Dusts causing pneumoconiosis generate OH$^{\bullet}$ and produce hemolisis by acting as Fenton catalysts. *Arch. Biochem. Biophys.*, 269, 359–364

Lakowicz, J.R., & Hylden, J.L. (1978) Asbestos-mediated membrane uptake of benzo(a)pyrene observed by fluorescence spectroscopy. *Nature*, 275, 446–448

Lapin, C.A., Graig, D.K., Valerio, M.G., McCandless, J.B. & Bogoroch, R. (1991) A subchronic inhalation study in rats exposed to silicon carbide whiskers. *Fundam. Appl. Toxicol.*, 16, 128–146

Leanderson, P. & Sahle, W. (1995) Formation of hydroxyl radicals and toxicity of tungsten oxide fibres. *Toxicol. In Vitro*, 9, 175–183

Leanderson, P., Söderkvist, P., Tagesson, C. & Axelson, O. (1988) Formation of 8-hydroxydeoxyguanosine by asbestos and man-made mineral fibres. *Br. J. Ind. Med.*, 45, 309–311

Leanderson, P., Söderkvist, P. & Tagesson, C. (1989) Hydroxyl radical mediated DNA base modification by manmade mineral fibres. *Br. J. Ind. Med.*, 46, 435–438

Levasseur-Acker, G., Zalma, R., Copin, E., Fournier, J., Pezerat, H. & Jankowski, R. (1995) Peroxidation de l'acide linolénique en présence des fibres d'amiante ou de némalite: résultats préliminaires avec des cellules épithéliales. *Can. J. Chem.*, 73, 453–459

Lison, D., Carbonnelle, P., Mollo, L., Lauwerys, R. & Fubini, B. (1995) Physicochemical mechanism of the interaction between cobalt metal and carbide particles to generate toxic activated oxygen species. *Chem. Res. Toxicol.*, 8, 600–606

Lu, L., Keane, M.J., Ong, T. & Wallace, W.E. (1994) In vitro genotoxicity studies of chrysotile asbestos fibers dispersed in simulated pulmonary surfactant. *Mutat. Res.*, 320, 253–259

Lund, L.G. & Aust, A.E. (1990) Iron mobilization from asbestos by chelators and ascorbic acid. *Arch. Biochem. Biophys.*, 278, 60–64

Lund, L.G. & Aust, A.E. (1992) Iron mobilization from crocidolite asbestos greatly enhances crocidolite-dependent formation of DNA single-strand breaks in FX174 RFI DNA. *Carcinogenesis*, 13, 637–642

Lund, L.G., Williams, M.G., Dodson, R.F. & Aust, A.E. (1994) Iron associated with asbestos bodies is responsible for the formation of single strand breaks in fX174RFI DNA. *Occup. Environ. Med.*, 51, 200–204

Mahmod, N., Khan, S.G., Ali, S., Athar, M. & Rahman, Q. (1993) Asbestos induced oxidative injury to DNA. *Ann. Occup. Hyg.*, 37, 315–319

Mahmod, N., Khan, S.G., Ali, S., Athar, M. & Rahman, Q. (1994) Differential role of hydrogen peroxide and organic peroxides in augmenting asbestos-mediated DNA damage: implications for asbestos induced carcinogenesis. *Biochem. Biophys. Res. Comm.*, 200, 687–694

Maples, K.R. & Johnson, N.E. (1992) Fiber-induced hydroxyl radical formation: correlation with mesothelioma induction in rats and humans. *Carcinogenesis*, 13, 2035–2039

Mason, R.P., Hanna, P.M., Burkitt, M.J. & Kadiiska M.B. (1994) Detection of oxygen-derived radicals in biological systems using electron spin resonance. *Environ. Health Perspectives*, 102, 33–36

McClellan, R.O., Miller, F.J., Hesterberger, T.W, Warheit, D.B., Bunn, W.B., Kane, A.B., Lippman, M., Mast, R.W., McConnell, E.E. & Reinhard, C.F. (1992) Approaches to evaluating the toxicity and carcinogenicity of man-made fibers: summary of a workshop held November 11–13, 1991, Durham, North Carolina. *Reg. Toxicol. Pharmacol.,* 16, 321–364

McFadden, D., Wright, J.L., Wiggs, B. & Churg, A. (1986) Smoking inhibits asbestos clearance. *Am. Rev. Respir. Dis.,* 133, 372–374

Mollo, L., Merlo, E., Giamello, E., Volante, M., Bolis, V. & Fubini, B. (1994) Effect of chelators on the surface properties of asbestos. In: Davis, J.M.G. & Jaurand, M.-C., eds, *Cellular and Molecular Effects of Mineral and Synthetic Dusts and Fibres* (NATO ASI Series Vol. H 85), Berlin, Springer-Verlag, pp. 425–432

Moyer, V.D., Cistulli, C.A., Vaslet, C.A. & Kane, A.B. (1994) Oxygen radicals and asbestos carcinogenicity *Environ. Health Perspectives*, 102 (Suppl. 10), 131–136

Mossman, B.T., Eastman A., Landesman, J.M. & Bresnick, E. (1983) Effects of crocidolite and chrysotile asbestos on cellular uptake and metabolism of benzo(*a*)pyrene in hamster tracheal epithelial cells. *Environ. Health Perspectives*, 51, 331–335

Mossman, B.T., Marsh, J.P., Sesko, A. Hill, S., Shatos, M.A., Doherty, J., Petruska, J., Adler, K.B., Hemenway, D., Mickey, R., Vacek, P. & Kagan, E. (1990) Inhibition of lung injury, inflammation and interstitial pulmonary fibrosis by polyethylene glyco-conjugated catalase in a rapid inhalation model of asbestosis. *Am. Rev. Respir. Dis.,* 41, 1266–1271

Nejjari, A., Fournier, J., Pezerat, H. & Leanderson, P. (1993) Mineral fibres: correlation between oxidizing surface activity and DNA base hydroxylation. *Br. J. Ind. Med.,* 50, 501–504

Pezerat, H., Zalma, R., Guignard, J. & Jaurand, M.C. (1989) Production of oxygen radicals by the reduction of oxygen arising from the surface activity of mineral fibers. In: Bignon, J., Peto, J. & Saracci, R., eds, *Non-Occupational Exposure to Mineral Fibres* (IARC Scientific Publications No. 90), Lyon, IARC, pp. 100–110

Pezerat, H., Guignard, J. & Cherrie, J.W. (1992) Man-made mineral fibers and lung cancer: an hypothesis. *J. Toxicol. Ind. Health,* 8, 77–87

Saracci, R. (1977) Asbestos and lung cancer: an analysis of the epidemiological evidence on the asbestos smoking interaction. *Int. J.Cancer,* 20, 323–331

Shatos, M.A., Doherty, J.M., Marsh, J.P. & Mossman, B.T. (1987) Prevention of asbestos-induced cell death in rat lung fibroblasts and alveolar macrophages by scavengers of active oxygen species. *Environ. Res.,* 44, 103–116

Shen, Z., Parker, V.D. & Aust, A.E. (1995) Mediated, thin-layer cell, coulometric determination of redox-active iron on the surface of asbestos fibers. *Anal. Chem.,* 67, 307–311

Stanton, M.F., Layard, M., Tegeris, A., Miller, E., May, M., Morgan, E., & Kent, E. (1977) Carcinogenicity of fibrous glass: pleural response in the rat in relation to fiber dimension. *J. Natl Cancer Inst.,* 58, 587–603

Thomas, G., Ando, T., Verma, K. & Kagan, E. (1994) Asbestos fibers and interferon-γ up-regulate nitric oxide production in rat alveolar macrophages. *Am. J. Respir. Cell Mol. Biol.,* 11, 707–715

Turver, C.J. & Brown, R.C. (1987) The role of catalytic iron in asbestos-induced lipid peroxidation and DNA-strand breakage in C3H10T1/2 cells. *Br. J. Cancer,* 56, 133–136

Vallyathan, V. (1994) Generation of oxygen radicals by minerals and its correlation to cytotoxicity. *Environ. Health Perspectives*, 102 (Suppl. 10), 111–115

Vallyathan, V., Mega, G.F., Shi, X. & Dalal, N.S. (1992) Enhanced generation of free radicals from phagocytes induced by mineral dusts. *Am. J. Respir. Cell Mol. Biol.,* 6, 404–413

Volante, M., Giamello, E., Merlo, E., Mollo, L. & Fubini, B. (1994) Surface reactivity of mechanically activated covalent solids and its relationship with the toxicity of freshly ground dusts. An EPR study. In: Tkacova, K. ed., *Proceedings of the 1st International Conference on Mechanochemistry*, Cambridge, Cambridge Interscience Publishing, pp. 125–130

Weitzman, S.A. & Graceffa, P. (1984) Asbestos catalyzes hydroxyl and superoxide radical generation from hydrogen peroxide. *Arch. Biochem. Biophys.,* 228, 267–274

Weitzman, S.A. & Weitberg, A.B. (1985) Asbestos-catalysed lipid peroxidation and its inhibition by desferroxamine. *Biochem. J.,* 225, 259–262

Zalma, R., Bonneau, L., Jaurand, M.-C., Guignard, J. & Pezerat, H. (1987a) Production of hydroxyl radicals by iron solid compounds. *Toxicol. Environ. Chem.,* 13, 171–188

Zalma, R., Bonneau, L., Jaurand, M.-C., Guignard, J. & Pezerat, H. (1987b) Formation of oxy-radicals by oxygen reduction arising from the surface activity of asbestos. *Can. J. Chem.,* 652, 338–2341

Corresponding author
B. Fubini
Dipartimento di Chimica Inorganica, Chimica Fisica e Chimica dei Materiali, Università di Torino, Torino, Italy

Use of in-vitro genotoxicity and cell transformation assays to evaluate the potential carcinogenicity of fibres

M.-C. Jaurand

Introduction

Neoplastic cell transformation results from molecular and cellular changes leading to autonomous proliferation. In-vitro genotoxicity and cell transformation assays are therefore important for the study of the mechanisms of fibre carcinogenicity. However, these systems have some limitations: in particular, cells *in vitro* are isolated systems and may not be relevant to the tissue of interest; in addition, cells *in vitro* can not be used to fully assess kinetic parameters such as clearance and biopersistence (McClellan et al., 1992), although some parameters of biopersistence, such as particle uptake and dissolution, can be evaluated.

Different types of in-vitro assays have been used to study the genotoxicity and transforming potential of fibres (for reviews, see Jaurand, 1989; Barrett et al., 1990; Brown et al., 1990; Jaurand, 1991; Walker et al., 1992).

Genotoxicity studies

Mutagenicity

The mutagenic potential of fibres has been investigated mainly by searching for revertants in bacteria and mutations at the *hprt* and HLA-A loci in mammalian cells (see Table 1). In early studies, either no mutagenic activity or only weak mutagenic activity was detected using various samples of asbestos and man-made vitreous fibres (MMVF), and it was concluded that prokaryotes may not be relevant systems for the study of the mutagenicity of fibrous matter. This conclusion was based on differences in membrane composition between prokaryotes and eukaryotes (from which differences in membrane–fibre interactions were assumed) and on the absence of fibre internalization in bacteria.

More recently, the mutagenicity of asbestos fibres was investigated with a new strain of bacteria, *Salmonella typhimurium* TA102, which is sensitive to oxygen radical DNA damage. Using this strain, Faux et al. (1994) found that UICC crocidolite, but not UICC chrysotile, produced the formation of revertants; Athanasiou et al. (1992), who also used this strain, did not observe mutations with tremolite.

While only a weak response was found with chrysotile or crocidolite at the *hprt* locus in A_L hamster–human cell hybrids, a significant rate of mutation was detected at the S1 locus (Hei et al., 1992). The S1 locus is located on the single human chromosome in the hybrid and the encoded genes are not necessary for survival. (The interpretation of results of mutation assays should take into account that if large mutations occur, loss of cell viability can impair the detection of mutants, depending on the importance for cell survival of the genes altered.) In this study, DNA analysis revealed large chromosomal deletions.

In human peripheral lymphocytes, fibre-induced mutations have been investigated in the autosomal HLA-A locus (Both et al., 1994). At doses of 50 µg/mL chrysotile was significantly mutagenic. However, no significant increase in mutation frequency was observed with 400 µg/mL crocidolite or erionite. In another HLA-A assay carried out with crocidolite, a mesothelioma cell line showed no significant increase in mutation frequency (Both et al., 1995).

In contrast to the absence of significant mutagenicity in the HLA-A assays, loss of heterozygosity was detected in human lymphocytes treated with crocidolite and erionite (in comparison with spontaneous mutants in control lymphocytes); chrysotile did not produce significant loss of heterozygosity (Both et al., 1994). A significant loss of heterozygosity was also observed with crocidolite in mesothelioma cells (Both et al., 1995).

These recent results indicate that some fibres do have mutagenic potential and can produce large deletions. However, the mechanisms whereby asbestos fibres produce these effects are unknown. Recent data indicate that fibre-induced mutagenesis may be induced by oxygen-derived molecules, as demonstrated by the protective effect of the antioxidant enzyme, manganese superoxide dismutase (SOD), in A_L cells (Hei et al., 1995).

Clastogenic effects
As shown in Table 2, asbestos fibres produce chromosomal damage in most in-vitro systems having as endpoints chromosome breakage (chromosome and chromatid breaks, fragments) and gaps and micronuclei formation.

In general, the levels of chromosomal damage (breaks and gaps) produced by asbestos remain moderate. For example, Kodama et al. (1993) observed a low but significant enhancement of chromosomal aberrations and micronuclei in human bronchial epithelial cells due to the cytogenetic effects of chrysotile; crocidolite produced an increase in micronuclei, but this effect was limited to a single incubation time. Pelin et al. (1995a) reported a low but significant enhancement of structural damage in two of six human mesothelial cell lines

Table 1. Studies on mutagenicity of asbestos fibres and MMVF[a]

System	Fibre types	Results	Reference
Salmonella typhimurium TA1538–TA1535 Escherichia coli, several strains	Chrysotile, crocidolite, amosite, anthophyllite MMVF (GF 100, GF 110)	No mutation	Chamberlain & Tarmy, 1977
Escherichia coli	Tremolite (Richterite)	With S9, mutation rate enhanced	Cleveland, 1984
Salmonella typhimurium TA102	Chrysotile, crocidolite (UICC samples)	Mutagenicity of crocidolite but not of chrysotile	Faux et al., 1994
Salmonella typhimurium TA102	Tremolite	No mutagenicity	Athanasiou et al., 1992
Chinese hamster lung cells	Chrysotile, crocidolite, amosite	Weak mutagenicity at the *hprt* locus	Huang, 1979
Adult rat liver cells	Chrysotile, crocidolite, amosite	No mutation at the *hprt* locus	Reiss et al., 1982
Syrian hamster embryo (SHE) cells	Chrysotile, crocidolite	No mutation at the *hprt* and Na^+/K^+ ATPase locus	Oshimura et al., 1984
Chinese hamster ovarian (CHO) cells	Crocidolite	No mutation at the *hprt* locus	Kenne et al., 1986
Hamster–human hybrid A_L	Chrysotile, crocidolite	Mutagenicity at the S1 locus. No mutagenicity at the *hprt* locus	Hei at al., 1992
Human lymphocytes (HLA-A)	Chrysotile, crocidolite, erionite	Mutagenicity of chrysotile (50 µg/mL); no mutagenicity of crocidolite or erionite (400 µg/mL); loss of heterozygosity HLA-A locus (crocidolite, erionite) positive	Both et al., 1994
Mesothelioma cell line (HLA-A)	Crocidolite	No mutagenicity; loss of heterozygosity (HLA-A locus) positive	Both et al., 1995

[a]MMVF, man-made vitreous fibres.

Table 2. In-vitro studies of chromosome damage by fibres

Cells[a]	Chrysotile	Crocidolite	Other fibres or particles[b]	Reference
SHE	+			Lavappa et al., 1975
SHE	+	+	+GF 100	Oshimura et al., 1984
SHE			+Tremolite	Athanasiou et al., 1992
SHE	+	+	+Amosite	Dopp et al., 1995
CHO	+			Babu et al., 1980
CHO		+	+Erionite, –Quartz	Kelsey et al., 1986
CHO[c]			+Amosite	Donaldson & Golyasnya, 1995
CHO-K1	+	+	–GF, –GP	Sincock & Seabright, 1975
CHO-K1	+	+	+GF 100, –GF 110	Sincock et al., 1982
CHL		+		Huang et al., 1978
CHLV79[c]	+			Lu et al., 1994
V79-4			–GF 110 (T), +GF 110 (R)	Brown et al., 1979
V79	+	+	+Erionite	Palekar et al., 1987
RTE	+[d]	–		Hesterberg et al., 1987
RPM	+			Jaurand et al., 1986
HBE	[+][e]	[+][e]		Kodama et al., 1993
HM			[+][e] Amosite[f]	Lechner et al., 1985
HM			+Amosite[g]	Pelin et al., 1995a
HM[h]	–	–	–Amosite[e]	Olofsson & Mark, 1989
HF	–	–	+GF 100, –GF 110	Sincock et al., 1982
HL	–	–	+GF 100, –GF 110	Sincock et al., 1982
HL	+	+		Valerio et al., 1983
HL[i]	+			Korkina et al., 1992

[a]Symbols: SHE, Syrian hamster embryo; CHO, Chinese hamster ovary; CHL, Chinese hamster lung; RTE, rat tracheal epithelial; RPM, rat pleural mesothelial; HBE, human bronchial epithelial; HM, human mesothelial; HF, human fibroblasts; HL, human lymphoid.
[b]Symbols: GF, glass fibre; GP, glass powder; T, total; R, respirable.
[c]Long fibres were more efficient than short fibres.
[d]Micronuclei.
[e]Weak effect.
[f]No breakage mentioned; dicentrics detected.
[g]Two of 10 primary cultures from 10 different donors gave significant results.
[h]No breaks; structural rearrangements (translocation, deletions) were found.
[i]Chromosome damage also observed with zeolite and latex particles.

from non-cancerous donors treated with amosite. In another study using amosite, no breaks but deletions and translocations were detected in human mesothelial cells (Olofsson & Mark, 1989). Chromosomal aberrations were observed in Syrian hamster embryo cells with one sample of tremolite (Athanasiou et al., 1992).

Chromosome breakage may be due to the formation of reactive oxygen/nitrogen species (ROS), as suggested by the studies on mutations. Korkina et al. (1992) found that structural chromosome aberrations in human lymphocytes were due to ROS, probably generated by monocytes present in the incubation medium. In the same way, the positive results obtained with lymphocytes and lymphoblastoid cell lines, summarized in Table 3, may be related to cooperation between monocytes and lymphocytes, since monocytes were not separated from lymphocytes in blood preparations. The hypothesis of a role of ROS in fibre carcinogenicity is confirmed by the occurrence of chromosome aberrations in human lymphocytes following

Table 3. Studies of chromosomal damage by fibres in lymphocytes and lymphoblastoid cell lines

Cell system	Fibres	Result	Reference
Human lymphoblastoid cell line, lymphocytes from ataxia telangiectasia	Crocidolite, chrysotile, GF 110	No polyploidy; no micronuclei	Sincock et al., 1982
Human lymphoblastoid cell line	Crocidolite, SFA chrysotile, GF 100	No increase in sister chromatid exchange	Casey, 1983
Human lymphocytes	Crocidolite, chrysotile	Chromatid and chromosome breaks	Valerio et al., 1983
Human lymphocytes (whole blood)	Chrysotile	Chromatid and chromosome breaks	Korkina et al., 1992
Human lymphocytes (whole blood)	Conditioned medium from chrysotile-treated RPMC	Enhancement of chromosome aberrations	Emerit et al., 1991

GF, glass fibres; RPMC, rat pleural mesothelial cells.

treatment with conditioned medium from chrysotile-treated mesothelial cells (Emerit et al., 1991). In this study, neither a cell-free nor an asbestos-free control-conditioned media produced chromosome damage. Also, the addition of catalase and SOD to the culture medium of asbestos-treated mesothelial cells reduced the clastogenic potency of the resultant conditioned medium.

Few data on clastogenicity have been reported with MMVF, although some positive results have been obtained in rodent cells (Table 2). Structural and numerical chromosomal aberrations have been reported in human mesothelial cells treated with thin glasswool and rockwool, with a stronger effect from samples containing the highest proportion of long thin fibres (Linnainmaa et al., 1991). Also, refractory ceramic fibres (RCF) produced nuclear abnormalities that were possibly related to chromosome damage in Chinese hamster ovarian (CHO) cells (Hart et al., 1992, 1994).

Mitotic abnormalities, aneuploidy and polyploidy
Anaphase aberrations have been observed in SHE cells and V79 cells treated with several types of fibres (Hesterberg & Barrett, 1985; Palekar et al., 1987; Dopp et al., 1995). Anaphase aberrations have also been studied in large SV40 T immortalized human mesothelial cells (MeT-5A); chrysotile and crocidolite significantly enhanced and amosite slightly enhanced the percentage of abnormal mitoses, whereas erionite did not (Pelin et al., 1992). A similar pattern was found in a series of assays using rat pleural mesothelial cells; those treated with chrysotile and crocidolite exhibited a significant enhancement in the frequency of anaphase/telophase abnormalities, whereas those treated with UICC amosite and MMVF samples did not (Yegles et al., 1993, 1995). However, although this result is in agreement with those obtained with human cells, the absence of an effect by amosite may be due to an insufficient number of 'critical' fibres in the amosite sample. This idea was suggested by Yegles et al. (1995) after studying the effects of 10 samples of asbestos fibres – mitotic abnormalities may not be detected unless a sufficient number of relevant fibres is reached, estimated to be 2×10^5 'Stanton' fibres per cm^2, a level that was reached in neither the amosite nor MMVF samples.

Aneuploidy and polyploidy have been frequently observed in fibre-treated cells (Table 4). This has been confirmed with SHE cells treated with tremolite (Athanasiou et al., 1992) but not in human bronchial epithelial cells treated with chrysotile and crocidolite (Kodama et al., 1993). The formation of binucleated cells has been used as an indicator of abnormal mitosis. Chrysotile and crocidolite induced an increase in the number of binucleated human bronchial epithelial cells (Kodama *et al.*, 1993), and UICC chrysotile and crocidolite and samples of glasswool and rockwool produced a significant increase in binucleation in primary and MeT-5A human mesothelial cells (Pelin *et al.*, 1995b). Rat liver epithelial cells seemed to be more sensitive in this respect than human cells to chrysotile (Pelin *et al.* 1995b). Bi- or multinucleated rat pleural mesothelial cells have been reported following treatment with chrysotile and crocidolite (Jaurand *et al.*, 1983).

While multinucleated cells can be observed after one round of replication, aneuploidy in metaphases is only observable after two rounds of replication. Thus, in assaying for aneuploidy, it is important that the time-point for evaluation is long enough for two complete cycles of cell division. Premature evaluation of aneuploidy may explain why human cells, which generally grow slowly, appear to have less fibre-induced aneuploidy than rodent cells, which tend to grow more rapidly.

Observations on the formation of numerical chromosome abnormalities (Lechner *et al.*, 1985), and the significant but low structural chromosome damage in some human mesothelial cell cultures (Pelin *et al.*, 1995a), suggest that interactions between

Table 4. In-vitro studies of aneuploidy and polyploidy in cells exposed to fibres (including binucleation)

Cells[a]	Chrysotile	Crocidolite	Other fibres or particles[b]	Reference
SHE	+	+	+GF 100	Oshimura *et al.*, 1984
SHE		+	+Tremolite	Athanasiou *et al.*, 1992
CHO			++Erionite, –MinUSil	Kelsey *et al.*, 1986
CHO		+		Kenne *et al.*, 1986
CHO	+	+	+RCF	Hart *et al.*, 1992
CHO-K1	+	+	[+][e] GF, –GP	Sincock & Seabright, 1975
CHO-K1	+	+	+GF 100, –GF 110	Sincock *et al.*, 1982
CHL		+		Huang *et al.*, 1978
V79-4			–MinUSil	Price-Jones *et al.*, 1980
V79	+	+	+Erionite	Palekar *et al.*, 1987
RTE	+	+		Hesterberg *et al.*, 1987
RTE	+	+	+GW, RW	Pelin *et al.*, 1995b
RPM	+[c]	+[c]		Jaurand *et al.*, 1983
RPM	+			Jaurand *et al.*, 1986
RPM		+		Yegles *et al.*, 1993
HBE	+[c,d]	[+][e]		Kodama *et al.*, 1993
HM			+Amosite	Lechner *et al.*, 1985
HM	+	+	+GW, RW	Pelin *et al.*, 1995b
HM[f]	+	+	+GW, RW	Pelin *et al.*, 1995b
HF	–	–	–GF 100, –GF 110	Sincock *et al.*, 1982
HF	+			Verschaeve & Palmer, 1985
HL	–	–	–GF 100, –GF 110	Sincock *et al.*, 1982
HL	–	–		Valerio *et al.*, 1983

[a]Symbols: SHE, Syrian hamster embryo; CHO, Chinese hamster ovary; CHL, Chinese hamster lung; RTE, rat tracheal epithelial; RPM, rat pleural mesothelial; HBE, human bronchial epithelial; HM, human mesothelial; HF, human fibroblasts; HL, human lymphoid.
[b]Symbols: GF, glass fibre; RCF, refractory ceramic fibres; GP, glass powder; GW, glasswool; RW, rockwool.
[c]Binucleation detected.
[d]No numerical change.
[e]Weak effect.
[f]MeT-5A (large SV40 T transfected).

Table 5. Studies on base hydroxylation produced by fibres

System[a]	Fibre	Results		Reference
1.3 mM calf thymus DNA, 2.5 mg/mL asbestos, 5 mM H_2O_2, 0.1 M PBS pH 7.4 with or without 5 mM EDTA, 20 h, 37°C, HPLC analysis	Crocidolite Amosite Anthophyllite Rhodesian chrysotile Control	8-OHdG/10^3 nt With EDTA 4.4 3.8 4.0 3.8 0.0	Without EDTA 10.8 16.6 4.3 7.2 0.0–4.2	Kasai & Nishimura, 1984
2 mg calf thymus DNA, 5 mg fibres, 60 mM PBS pH 7.4, 3 h, 37°C, HPLC analysis	Crocidolite	PBS alone 0.5 mM Fe_2O_3 0.5 mM H_2O_2 + 0.5 mM Fe_2O_3 + 1 mM EDTA	8-OHdG/10^5 dG[b] 15 15 336 375 4315	Adachi et al., 1994
	De-ironized crocidolite(HCl andEDTA treated)	PBS alone 0.5 mM Fe_2O_3 0.5 mM H_2O_2 +0.5 mM Fe_2O_3 + 1 mM EDTA	30 18 576 936 6916	
	Fibre-free control	PBS alone 0.5 mM Fe_2O_3 0.5 mM H_2O_2 + 0.5 mM Fe_2O_3 + 1 mM EDTA	8 21 383 972 180	
1 mg/mL calf thymus DNA, 2.5 mg/mL fibres, 0.1 M PBS pH 7.4, 20 h, 37°C, HPLC analysis	Asbestos[c]	No additive 5 mM H_2O_2 0.5 mM H_2O_2 0.5 mM H_2O_2 + 1mM EDTA	8-OHdG/10^5 dG enhancement over control +3 to 100 + 300 to 1200 +100 to 500 +600 to 1500	Adachi et al., 1992
	MMVF[c]	No additive 5 mM H_2O_2 0.5 mM H_2O_2 0.5 mM H_2O_2 + 1 mM EDTA	−7 and −2 −140 and −80 −70 and −25 −2 and +115	
0.5 mg/mL calf thymus DNA, 10 mg/mL fibres, PBS, 5 h, 37°C, HPLC analysis	MMVF (from Rockwool Inc.)	No fibre Fibres alone + 100 μM H_2O_2 + 1 mM H_2O_2 + catalase + 10 mM DMSO	8-OGdG10^5 dG 5.1 4.2 ± 0.2[d] 11.3 ± 1.3[d] 26.5 ± 1.1[d] 0.2 ± 0.3[d] 2.0 ± 0.5[d]	Leanderson et al., 1989

Table 5 contd. Studies on base hydroxylation produced by fibres

System[a]	Fibre	Results		Reference
500 µg/mL calf thymus DNA, 5 mg/mL fibres, 0.1 M KH_2PO_4 buffer pH 7.4, 5 h, 37°C, HPLC analysis	Crocidolite	No additive 1 mM H_2O_2 fibres + 1 mM desferrioxamine + 1 mM H_2O_2	Percentage 8-OHdG/dG 0.01 0.07 0.10 0.07 0.22	Faux et al., 1994
DNA extraction from HL60 cells treated with fibres, incubation for 20 h, HPLC analysis	Crocidolite	 15 µm/mL 50 µm/mL 150 µm/mL	8-OHdG /10^5dG enhancement over control (max. increase 12–20 h) 1.0 1.5 2.0	Takeuchi & Morimoto, 1994

[a]EDTA, ethylene diamine tetra-acetic acid; HPLC, high-performance liquid chromatography; PBS, phosphate buffer solution; DMSO, dimethyl sulfoxide.
[b]Mean of three independent samples of the same incubation.
[c]Asbestos: crocidolite, amosite and Rhodesian and Canadian chrysotile. MMVF, glasswool, potassium titanate whiskers.
[d]Mean ± standard deviation of four experiments.

human mesothelial cells and amosite fibres do not result in the production of great amounts of clastogenic factors. This is in agreement with the failure to detect ROS production in human mesothelial cells treated with this fibre (Gabrielson et al., 1986; Kinnula et al., 1994).

DNA damage

That chromosome changes are observed in different cell types treated with asbestos fibres, and to a lesser extent with MMVF, indicates that mineral fibres produce DNA damage. Direct DNA breakage has been poorly investigated and no direct demonstration of DNA strand breaks has been made. For example, Kinnula et al. (1994) reported no single-strand breaks in MeT-5A cells treated with amosite. However, DNA breakage has been shown indirectly by the occurrence of (i) DNA repair in rat pleural mesothelial cells (Renier et al., 1990); (ii) nick translation in rat embryo cells treated with asbestos (Libbus et al., 1989); and (iii) significant numbers of DNA nicks in C3H10T1/2 cells produced by both native and milled asbestos (Turver & Brown, 1987). No DNA repair was observed in hepatocytes following treatment with chrysotile (Denizeau et al., 1985) but erionite enhanced DNA repair in C3H10T1/2 cells (Poole et al., 1983).

Several key processes control DNA integrity. Poly(ADP)ribosylation is induced by DNA breaks and is involved in the repair of damage. Poly(ADP)ribosylation has been observed following the treatment of rat pleural mesothelial cells with chrysotile and crocidolite (Dong et al., 1995), and, through the use of poly(ADP)ribose polymerase inhibitors, it was found that fibre cytotoxicity was not the result of DNA damage.

The observation that ROS are produced during fibre–cell interactions has led to the study of DNA base adduct formation by asbestos fibres and MMVF. There is a natural background level of 8-hydroxydeoxyguanosine (8-OHdG) that is estimated to be approximately 0.5–2.0 8-OHdG/10^5 guanine, and although higher levels of 8-OHdG can be found in normal tissues, these basal levels are enhanced following treatment with carcinogens (Floyd, 1990). The investigation of guanine hydroxylation has been carried out mainly in vitro, typically in acellular systems that use calf thymus DNA or deoxyguanosine (dG) as a substrate (see Table 5). Asbestos fibres, including amosite, have been found to produce an enhancement

Table 6. Studies of cell transformation by fibres

Cell system	Fibres	Assay	Results	Reference
BALB/3T3	Chrysotile, crocidolite	Foci formation	Chrysotile: initiator and complete carcinogen; crocidolite: initiating-like effects	Lu et al., 1988
C3H 10T1/2	Chrysotile, crocidolite	Foci formation	No transformation by asbestos; transformation by asbestos after γ-irradiation	Hei et al., 1984
SHE	Chrysotile, crocidolite, amosite, anthophyllite	Colony formation	Low transformation; enhancement of transformation with benzo[a]pyrene	DiPaolo et al., 1983
SHE	Chrysotile, crocidolite, GF 100, GF 110	Colony formation	Transformation with asbestos and glass fibres	Hesterberg & Barrett, 1984
SHE	Chrysotile, crocidolite, amosite, anthophyllite, GF 100, GF 110	Colony formation	Transformation with asbestos and GF 100; crocidolite: initiator	Mikalsen et al., 1988
SHE	Tremolite	Colony formation	Transformation	Athanasiou et al., 1992
Rat pleural mesothelial cells	Chrysotile	Colony formation	Transformation: abnormal colonies in liquid medium, growth in semi-solid medium and tumorigenesis in nude mice	Patérour et al., 1985; Saint-Etienne et al., 1993

SHE, Syrian hamster embryo; GF, glass fibre.

of 8-OHdG in calf thymus DNA after incubation for several hours in phosphate buffer solution at pH 7.4 (Adachi et al., 1992, 1994; Faux et al. 1994). This enhancement over control was generally potentiated by the addition of hydrogen peroxide and was enhanced further in the presence of ethylene diamine tetra-acetic acid (EDTA) (Leanderson et al., 1988, 1989; Adachi et al., 1992, 1994; Takeuchi & Morimoto, 1994). It should also be noted that de-ironized crocidolite also produced guanine hydroxylation in the presence of hydrogen peroxide and EDTA. MMVF samples tested so far also were also found to enhance the formation of adducts (Leanderson et al., 1988, 1989; Adachi et al., 1992). Few studies have been carried out using cell systems. Takeuchi and Morimoto (1994) found an increase in 8-OHdG in the DNA of HL60 cells treated with crocidolite; they estimated that 2 in 10^5 dG were altered to 8-OHdG.

Cell transformation

Cell transformation assays are based on the determination of morphological changes in cells *in vitro*. In comparison with mutation assays, which are indicative mainly of single changes, cell transformation assays provide evidence for effects on several stages in neoplastic progression. Transformed cells can be detected when they form colonies (foci) of abnormal cells in confluent cultures. Few data have been published using transformation assays to investigate the carcinogenicity of fibres (see Table 6). Tests have been carried out using mainly SHE cells, C3H10T1/2 and rat pleural mesothelial cells using asbestos fibres; transformation by MMVF has been tested using SHE cells. Thin glasswool fibres (of 0.13 µm diameter) were found to be more cytotoxic and transforming than thicker glasswool fibres (diameter, 0.8 µm) on a weight-for-weight basis (Hesterberg & Barrett, 1984). Using the same transformation assay, Hesterberg et al. (1986) demonstrated that long glasswool fibres are more active than short fibres (milled long fibres) and that the reduction in cell transformation was not related to a reduced number of fibres internalized. Rat pleural mesothelial cells have been transformed by chrysotile fibres in a long-term assay (Patérour et al., 1985; Saint-Etienne et al., 1993).

Inhibition of intercellular communication

Mesothelial cells exhibit gap junctional intercellular communication (GJIC) (Pelin et al., 1994). GJIC is typically reduced by agents promoting carcinogenesis. For example, 12-O-tetradecanoyl phorbol acetate (TPA), a tumour promoter, decreases GJIC in normal mesothelial cells and inhibits metabolic cooperation in V79 and SHE cells (Table 7). Asbestos fibres and MMVF (glasswool, coarse glass) tested in these assays did not significantly change GJIC or metabolic cooperation.

Gene expression and cell proliferation

Asbestos fibres and MMVF have been found to stimulate the expression of several genes in mesothelial and/or epithelial cells. These genes include those that code for protective enzymes (e.g. SOD), ornithine decarboxylase and transcription or related factors (e.g. c-*fos* and c-*jun*) (Mossman et al., 1986; Marsh & Mossman, 1991; Heintz et al., 1993; Janssen et al., 1994a,b). Some of these genes also play a role in the control of the cell proliferation.

Conflicting data have been reported regarding the effect of fibres on cell proliferation – both the stimulation and inhibition of cell proliferation has been reported (Table 8). However, cell proliferation is regulated by many factors, including both negative and positive growth regulators and survival factors. Progression through the cell cycle is regulated by a complex sequence of phosphorylation/dephosphorylation reactions by the cyclin/CDK complexes and phosphatases and the activation/inactivation of transcription factors that may induce or repress specific genes. Table 8 summarizes the results obtained with different cell types; it is not intended to be exhaustive but illustrates some results with different systems; there are notable differences in experimental design.

Lasky et al. (1995) reported that different culture conditions can either enhance or decrease fibroblast proliferation after a three-day treatment with asbestos in serum-free conditions.

Table 7. Studies of cell communication and metabolic cooperation in cells exposed to fibres

Cell type	Fibres	Effect on communication[a]/ metabolic cooperation[b]	Reference
Normal mesothelial cells	100 ng/mL TPA[c]	Decrease in GJIC	Linnainmaa et al., 1991
	1–2 µg/cm^2 amosite	No decrease in GJIC	
	1 µg/cm^2 chrysotile	No decrease in GJIC	
	2 µg/cm^2 thin glasswool	No decrease in GJIC	
	1 µg/cm^2 coarse glasswool	No decrease in GJIC	
V79-4 Chinese hamster lung cells	0.01 µg/mL TPA	Inhibition in metabolic cooperation	
	0.01–5.0 µg/mL amosite	No inhibition in metabolic cooperation	
	0.01–5.0 µg/mL milled amosite	No inhibition in metabolic cooperation	Chamberlain, 1982
V79 or BPNi cells	0.3–4 µg/cm^2 tremolite	No effect on GJIC	Athanasiou et al., 1992

[a]Communication: GJIC, gap junctional intercellular communication.
[b]Metabolic cooperation: cloning efficiency of 8-azaguanine-resistant cells in the presence of 8-azaguanine-sensitive cells.
[c]TPA, 12-O-tetradecanoyl phorbol acetate (positive control).

Reduction in cell proliferation can follow cell injury, especially DNA damage, and the stimulation of cell proliferation can be a later event associated with repair after cell damage. The reduction of cell proliferation may be due to cell death in certain conditions, especially at high fibre concentrations; however, such a reduction is also observed sometimes without or with only minor cell death (Kodama et al., 1993; Hart et al., 1994), and it is this latter cell-cycle arrest without associated cell death that represents the regulated inhibition of cell proliferation following DNA damage. In CHO cells, a low level of oxidative stress blocks cell proliferation (Clopton & Saltman, 1995), and it is known that ROS are produced as a result of cell–fibre interactions in phagocytic cells (Kamp et al., 1992) and in other cell types able to phagocytize fibres (Mossman & Landesman, 1983;

Table 8. Studies of in-vitro stimulation of cell proliferation by fibres

System	Exposure	Assay	Result	Reference
Fibroblasts	UICC asbestos	Cell proliferation: ^{3}HdThd incorporation	Reduction after 24 h and resumption at 48 h	Lemaire et al., 1982
Hamster tracheal organ cultures	Asbestos, MMVF	^{3}HdThd labelling	Enhanced labelling 2 weeks after exposure with long chrysotile (not with short), crocidolite and GF 100	Woodworth et al., 1983
Rat tracheal epithelial cells	Chrysotile, crocidolite	Colony formation	Reduction of cell proliferation	Hesterberg et al., 1987
Rat primary fibroblasts	Chrysotile, crocidolite, amosite, anthophyllite	Mitotic index	Decrease in mitotic index	Wydler et al., 1988
Hamster tracheal epithelial cells	X–XO, H_2O_2 Asbestos	^{3}HdThd incorporation in culture medium supplemented with 10% FBS	24 h: no significant enhancement or decrease; 48 h: enhancement at medium doses	Marsh & Mossman, 1991
Hamster tracheal epithelial cells	Asbestos	Serum deprivation 24 h before addition of dusts	Cell proliferation	Janssen et al., 1994a
Human bronchial epithelial cells	Chrysotile, crocidolite	Colony formation	Inhibition of cell proliferation	Kodama et al., 1993
Rat pleural mesothelial cells	Chrysotile, crocidolite, amosite	^{3}HdThd incorporation in proliferative cells treated with/without asbestos	Decrease in ^{3}HdThd incorporation with asbestos	Dong et al., 1994
Chinese hamster ovary cells	Asbestos, RCF	Colony formation (RCF); cell count (all fibres)	Inhibition of cell proliferation	Hart et al., 1992, 1994
Human mesothelial cells	Amosite	Mitotic index in semiconfluent cultures	1 µg/cm^2: +0.9% ($n = 7$); –3.3% ($n = 3$) 2 µg/cm^2: +3% ($n = 1$); –3.1% ($n = 9$)	Pelin et al., 1995a

UICC, Union Internationale Contre le Cancer; ^{3}HdThd, ^{3}H-deoxythymidine; MMVF, man-made vitreous fibre; GF, glass fibre; FBS, fetal bovine serum; RCF, refractory ceramic fibres.

Table 9. Studies of in-vivo stimulation of cell proliferation by fibres

Exposure/system	Assay	Result	Reference
Inhalation of chrysotile in rat	³HdThd incorporation in airway epithelial cells	33 h post exposure: enhancement of ³HdThd incorporation	McGravan & Brody, 1989
Intratracheal instillation of long/short amosite in rat	Kinetics of BrdU incorporation in pulmonary and mesothelial cells	Greater incorporation with long fibres; enhancement of incorporation after one day (pulmonary cells), 3 days (mesothelial cells)	Adamson et al., 1993
Inhalation of amosite and MMVF in rats	BrdU uptake in lung cells	16 h post exposure: enhancement in pulmonary cells with amosite – not with MMVF	Donaldson et al., 1995
Injection of crocidolite in the peritoneal cavity of mice	DNA labelling	Diaphragmatic mesothelial cell proliferation 3 days after inoculation	Moalli et al., 1987

³HdThd, ³H-deoxythymidine; BrdU, 5-bromo-2´deoxyuridine; MMVF, man-made vitreous fibres.

Garcia et al., 1988; Dong et al., 1994). It should be noted that the cell proliferation observed *in vivo* after inhalation or intratracheal instillation of becomes evident only after some time (Table 9). For example, proliferation of peritoneal mesothelial cells has been observed 3–7 days after fibre inoculation (Moalli et al., 1987; Moyer et al., 1994).

The stimulation of gene expression by asbestos fibres can be due to either the triggering of events involved in repair of DNA damage or the stimulation of cell proliferation. For instance, c-*fos* is induced by DNA-damaging molecules. Interestingly, the induction of c-*fos* in JB6 cells (mouse epidermal cells) was not associated with the activation of poly(ADP)ribose polymerase (Amstadt et al., 1992), as has been observed in cultures of mesothelial cells (Dong et al., 1995).

Discussion

Although conventional point mutation assays (e.g. the Ames test and the mammalian *hprt* assay) have not shown increased mutation rates associated with fibres, assays measuring global DNA damage or large deletions, numerical or structural chromosome damage and mitotic disturbances have shown mainly positive responses. These latter assays thus appear to be useful for the testing of fibre carcinogenicity; asbestos fibres should produce mutations in model systems that are sensitive to the effects of ROS and are able to detect large deletions, at least in eukaryotic cells.

The extent of mutagenesis may depend on a cell's ability to adapt to oxidant stress. In this context, an induction of SOD and haeme oxygenase (HO) in epithelial and/or mesothelial cells treated with chrysotile and crocidolite has been demonstrated (Mossman et al, 1986; Janssen et al. 1994b); this is in agreement with the observation that hydrogen peroxide produces DNA strand breaks in epithelial and mesothelial cells (Kinnula et al. 1994; Churg et al., 1995). HO was also induced in A_L cells treated with chrysotile fibres (Suzuki & Hei, 1996). Although antioxidant enzymes added to the culture medium were found to protect against chromosome and DNA damage (Korkina et al., 1992; Dong et al., 1994; Hei et al., 1995), the addition of antioxidants acting at the intracellular level does not reduce genotoxicity (Suzuki & Hei, 1996). This finding may indicate that ROS are produced in the extracellular medium and that secondary messengers are responsible for any damage to genetic material.

Chromosome damage produced by fibres, especially in terms of chromosomal missegregation and aneuploidy, is likely to be an important step in asbestos-induced cell transformation. The involvement of ROS in anaphase aberrations remains to be demonstrated and was not evident with rat pleural mesothelial cells treated with crocidolite (Yegles et al.,

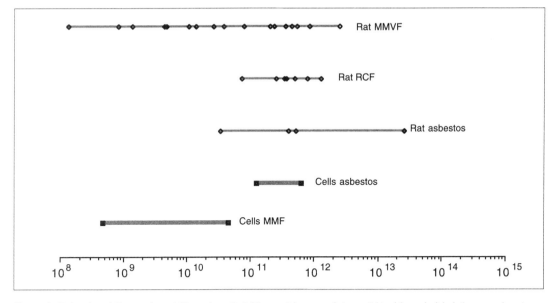

Figure 1. Estimation of the number of fibres deposited (diamonds) per g of dry weight of lung, in inhalation experiments, or deposited (squares) per g of cell proteins, in in-vitro experiments. The amount of fibres inhaled in inhalation experiments was calculated from the cumulative exposure ($F \times mL^{-1} \times h^{-1}$) multiplied by the volume of air inhaled during the duration of exposure, and deposition was assumed to be approximately 10% of the amount inhaled. In-vitro deposition was extrapolated from the protein weight in cells; exposure was considered to be 1–10 µg/cm² asbestos and 1–100 µg/cm² MMVF. Data on rat exposure (nose-only) are from Mast et al. (1995a,b), McConnell (1994) and Smith et al. (1987).

1995). The physical interaction between fibres and the cytoskeleton (Ault et al., 1995) might result in chromosomal missegregation. The cytogenetic response of human mesothelial cells to amosite, determined by chromosomal gaps and breaks, did not depend on the glutathione-S-transferase genotype of the mesothelial cell donor. However, cells lacking GSTM1 were more sensitive to the toxic effects of these fibres than cells with this gene (Pelin et al., 1995b).

By their nature, ROS are short-lived molecules; however, evidence from systems other than cell–fibre interactions has shown that ROS can produce more stable clastogenic factors (Emerit et al. 1995), which, if produced in response to fibres, could cause damage remote from the site of production. Indeed, clastogenic factors may be produced in association with particle internalization, as suggested by the following data: (i) phagocytosis is inhibited by antioxidants (Hobson et al., 1990) and (ii) 8-OHdG formation in DNA of differentiated HL60 cells is dependent on phagocytosis (Takeuchi & Morimoto, 1994). However, it is difficult to distinguish between radicals originating from cell–fibre interactions or from culture medium–fibre interactions.

As noted above, the constitutive iron in crocidolite asbestos may be mobilized and removed. Intracellular release of iron might be associated with cytotoxicity (Chao et al., 1994). However, Fe(III) at the fibre surface may be reduced by lung fluid and cells, an effect that would change the surface reactivity of the fibres (Ghio et al., 1994) and that may be of importance in vivo.

From these in-vitro studies of mutagenicity and clastogenicity, it seems that the interaction between cells and fibres can result in genetic damage. With asbestos fibres, these findings are in agreement with the results of transformation assays and animal experiments, at least for chrysotile and crocidolite, the only fibre types tested in these different model systems. Insufficient data are available on the effects of other types of asbestos, such as amosite and tremolite; it is particularly important that the mutagenic potency of tremolite is evaluated – preferential retention of tremolite has been noted in human lungs, in comparison with chrysotile, and tremolite is frequently associated with asbestos dusts. In addition, it is important to evaluate the mutagenic potency of fibre samples of different size. Finally, few data are available at present on the genotoxicity and transforming potency of MMVF.

Cell transformation assays have demonstrated a greater carcinogenicity of long versus short fibres, which is in good agreement with the results of animal experiments using injection and inhalation. However, it must be stressed that different samples of the same fibre type should be tested, since the response can differ significantly with samples of different origins (Jaurand et al., 1987; Yegles et al., 1995). Moreover, it is essential to determine, for each of the different in-vitro assays, the minimal amount of fibre needed to detect a significant effect. The relative importance of physico-chemical properties of fibres (e.g. dimensions and generation of ROS) in the induction of DNA damage, mutagenic effects and/or cell transformation must also be determined.

The in-vitro experimental results obtained with crocidolite and chrysotile do not emphasize major differences in the carcinogenic potency of these fibre types, in spite of some differences in fibre potency depending on the method used to measure it (number of fibres versus weight). Comparisons have to be made with caution because different samples will differ in their size distributions and number of fibres per unit weight. For this reason it is almost impossible to make stringent comparisons of the effects of one sample of a mineral fibre with another. Animal studies have suggested that carcinogenic potency differs depending on the dimensions of fibres.

Even less straightforward than comparisons between different fibre samples are comparisons of exposure in vivo and in vitro. In humans and rats, the lung burden is generally expressed in terms of dry weight. For comparison with in-vivo data, in-vitro exposure can be expressed with respect to cell mass and to the total number of fibres added to the culture. Figure 1 represents an attempt to compare these different experimental settings; it shows that the amount of fibres applied to in-vitro cell systems remains in the range of that used and assumed to be deposited in the lungs in inhalation experiments in animals. It can be noted that the 'cumulative exposures' of the cells treated with MMVF are much lower than that of asbestos.

In conclusion, the in-vitro models are useful in the investigation of the carcinogenic potency of fibres. The results obtained with asbestos emphasize the utility of the in-vitro studies. The development of assays based on analyses of chromosome disturbances and DNA damage seem appropriate to the investigation of the mechanistic effects of fibres. To allow comparisons of different fibre samples, genotoxicity assays and transformation studies should be conducted using a variety of mineral fibres and MMVF samples. To compare in-vitro and animal findings, it important to use similar samples and, as far as possible, samples representative of human exposures.

References

Adachi, S., Kawamura, K., Yoshida, S. & Takemoto, K. (1992) Oxidative damage on DNA induced by asbestos and man-made fibers in vitro. *Int. Arch. Occup. Environ. Health*, 63, 553–557

Adachi, S., Yoshida, S., Kawamura, K., Takahashi, M., Uchida, H., Odagiri, Y. & Takemoto, K. (1994) Inductions of oxidative DNA damage and mesothelioma by crocidolite, with special reference to the presence of iron inside and outside of asbestos fiber. *Carcinogenesis*, 15, 753–758

Adamson, I.Y.R., Bakowska, J. & Bowden, D.H. (1993) Mesothelial cell proliferation after instillation of long or short asbestos fibers into mouse lung. *Am. J. Pathol.*, 142, 1209–1216

Amstad, P.A., Krupitza, G. & Cerutti, P.A. (1992) Mechanism of c-fos induction by active oxygen. *Cancer Res.*, 52, 3952–3960

Athanasiou, K., Constantopoulos, S. H., Rivedal, E., Fitzgerald, D. J. & Yamasaki, H. (1992) Metsovo-tremolite asbestos fibres: in vitro effects on mutation, chromosome aberrations, cell transformation and intercellular communication. *Mutagenesis*, 7, 343–347

Ault, J.G., Cole, R.W., Jensen, C.G., Jensen, L.C.W., Bachert, L.A. & Rieder, C.L. (1995) Behavior of crocidolite asbestos during mitosis in living vertebrate lung epithelial cells. *Cancer Res.*, 55, 792–798

Babu, K.A., Lakkad, B.C., Nigam, S.K., Bhatt, D.K., Karnik, A.B., Thakore, K.N., Kashyap, S.K. & Chatterjee, S.K. (1980) In vitro cytological and cytogenetic effects of an Indian variety of chrysotile asbestos. *Environ. Res.*, 21, 416–422

Barrett, J.C., Lamb, P.W. & Wiseman, R.W. (1990) Hypotheses on the mechanisms of carcinogenesis and cell transformation by asbestos and other mineral dusts. *NATO ASI Series*, Vol. G21, 282–307

Both, K., Henderson, D.W. & Turner, D.R. (1994) Asbestos and erionite fibres can induce mutations in human lymphocytes that result in loss of heterozygosity. *Int. J. Cancer*, 59, 538–542

Both, K., Turner, D.R. & Henderson, D.W. (1995) Loss of heterozygosity in asbestos-induced mutations in a human mesothelioma cell line. *Environ. Mol. Mutag.*, 26, 67–71

Brown, R.C., Chamberlain, M., Davies, R., Gaffen, J. & Skidmore, J.W. (1979) In vivo biological effects of glass fibres. *J. Environ. Pathol. Toxicol.*, 2, 1369–1383

Brown, R.C., Hoskins, J.A., Miller, K. & Mossman, B.T. (1990) Pathogenic mechanisms of asbestos and other mineral fibres. *Mol. Aspects Med.*, 11, 35–349

Casey, G. (1983) Sister-chromatid exchange and cell kinetics in CHO-K1 cells, human fibroblasts and lymphoblastoid cells exposed in vitro to asbestos and glass fibre. *Mutat. Res.*, 116, 193–377

Chamberlain, M. (1982) The influence of mineral dusts on metabolic co-operation between mammalian cells in tissue culture. *Carcinogenesis*, 3, 337–339

Chamberlain, M. & Tarmy, E.M. (1977) Asbestos and glass fibres in bacterial mutation tests. *Mutat. Res.*, 43, 159–164

Chao, C.C., Lund, L.G., Zinn, K.R. & Aust, A.E. (1994) Iron mobilization from crocidolite asbestos by human lung carcinoma cells. *Arch. Biochem. Biophys.*, 314, 384–391

Churg, A., Keeling, B., Gilks, B., Porter, S. & Olive, P. (1995) Rat mesothelial and tracheal epithelial cells show equal DNA sensitivity to hydrogen peroxide-induced oxidant injury. *Am. J. Physiol.*, 268, L832–L838

Cleveland, M. G. (1984) Mutagenesis of Escherichia coli (CSH50) by asbestos (41954). *Proc. Soc. Exp. Biol. Med.*, 177, 343–346

Clopton, D.A. & Saltman, P. (1995) Low-level oxidative stress causes cell-cycle specific arrest in cultured cells. *Biochem. Biophys. Res. Commun.*, 210, 189–196

Denizeau, F., Marion, M., Chevalier, G. & Cote, M. (1985) Inability of chrysotile asbestos fibers to modulate the 2-acetylaminofluorene-induced UDS in primary cultures of hepatocytes. *Mutat. Res.*, 155, 83–90

DiPaolo, J.A., DeMarinis, A.J. & Doniger, J. (1983) Asbestos and benzo(*a*)pyrene synergism in the transformation of syrian hamster embryo cells. *Pharmacology*, 27, 65–73

Donaldson, K. & Golyasnya, N. (1995) Cytogenetic and pathogenic effects of long and short amosite asbestos. *J. Pathol.*, 177, 303–307

Donaldson, K., Brown, D.M., Miller, B.G. & Brody, A.R. (1995) Bromo-deoxyuridine (BRDU) uptake in the lungs of rats inhaling amosite asbestos or vitreous fibres at equal airborne fibre concentrations. *Exp. Toxicol. Pathol.*, 47, 207–211

Dong, H.Y., Buard, A., Renier, A., Levy, F., Saint-Etienne, L. & Jaurand, M.-C. (1994) Role of oxygen derivatives in the cytotoxicity and DNA damage produced by asbestos on rat pleural mesothelial cells in vivo. *Carcinogenesis*, 15, 1251–1255

Dong, H.Y., Buard, A., Levy, F., Renier, A., Laval, F. & Jaurand, M.-C. (1995) Synthesis of poly(ADP-ribose) in asbestos treated rat pleural mesothelial cells in culture. *Mutat. Res.*, 331, 197–204

Dopp, E., Saedler, J., Stopper, H., Weiss, D.G. & Schiffmann, D. (1995) Mitotic disturbances and micronucleus induction in Syrian hamster embryo fibroblast cells caused by asbestos fibers. *Environ. Health Perspectives*, 103, 268–271

Emerit, I., Jaurand, M.-C., Saint-Etienne, L. & Levy, F. (1991) Formation of a clastogenic factor by asbestos-treated rat pleural mesothelial cells. *Agents Actions*, 34, 410–415

Emerit, I., Levy, A., Pagano, G., Pinto, L., Calzone, R. & Zatterale, A. (1995) Transferable clastogenic activity in plasma from patients with Fanconi anemia. *Hum. Genet.*, 96, 14–20

Faux, S.P., Howden, P.J. & Levy, L.S. (1994) Iron-dependent formation of 8-hydroxydeoxyguanosine in isolated DNA and mutagenicity in *Salmonella typhimurium* TA102 induced by crocidolite. *Carcinogenesis*, 15, 1749–1751

Floyd, R.A. (1990) The role of 8-hydroxyguanine in carcinogenesis. *Carcinogenesis*, 11, 1447–1450

Gabrielson, E.W., Rosen, G.M., Grafstrom, C., Strauss, K.E. & Harris, C.C. (1986) Studies on the role of oxygen radicals in asbestos-induced cytopathology of cultured human lung mesothelial cells. *Carcinogenesis*, 7, 1161–1164

Garcia, J.G.N., Gray, L.D., Dodson, R.F. & Callahan, K.S. (1988) Asbestos-induced endothelial cell activation and injury. *Am. Rev. Respir. Dis.*, 138, 958–964

Ghio, A.J., Kennedy, T.P., Stonehuerner, J.G., Crumbliss, A.L. & Hoidal, J.R. (1994) DNA strand breaks following in vitro exposure to asbestos increase with surface-complexed [Fe^{3+}]. *Arch. Biochem. Biophys.*, 311, 13–18

Hart, G.A., Kathman, L.M. & Hesterberg, T.W. (1994) In vitro cytotoxicity of asbestos and man-made vitreous fibers: roles of fiber length, diameter and composition. *Carcinogenesis*, 15, 971–977

Hart, G.A., Newmen, M.M., Bunn, W.B. & Hesterberg, T.W. (1992) Cytotoxicity of refractory ceramic fibres to chinese hamster ovary cells in culture. *Toxicol. In Vitro*, 6, 317–326

Hei, T.K., Hall, E.J. & Osmak, R.S. (1984) Asbestos, radiation and oncogenic transformation. *Br. J. Cancer*, 50, 717–720

Hei, T.K., Piao, C.Q., He, Z.Y., Vannais, D. & Waldren, C.A. (1992) Chrysotile fiber is a strong mutagen in mammalian cells. *Cancer Res.*, 52, 6305–6309

Hei, T.K., He, Z.Y. & Suzuki, K. (1995) Effects of antioxidants on fiber mutagenesis. *Carcinogenesis*, 16, 1573–1578

Heintz, N.H., Janssen, Y.M. & Mossman, B.T. (1993) Persistent induction of c-fos and c-jun expression by asbestos. *Proc. Natl Acad. Sci. USA*, 90, 3299–3303

Hesterberg, T.W. & Barrett, J.C. (1984) Dependence of asbestos- and mineral dust-induced transformation of mammalian cells in culture on fiber dimension. *Cancer Res.*, 44, 2170–2180

Hesterberg, T.W. & Barrett, J.C. (1985) Induction by asbestos fibers of anaphase abnormalities: mechanism for aneuploidy induction and possibly carcinogenesis. *Carcinogenesis*, 6, 473–475

Hesterberg, T.W., Butterick, C.J., Oshimura, M., Brody, A.R. & Barrett, J.C. (1986) Role of phagocytosis in Syrian hamster cell transformation and cytogenetic effects induced by asbestos and short and long glass fibers. *Cancer Res.*, 46, 5795–5802

Hesterberg, T.W., Ririe, D.G., Barrett, J.C. & Nettesheim, P. (1987) Mechanisms of cytotoxicity of asbestos fibres in rat tracheal epithelial cells in culture. *Toxicol. In Vitro*, 1, 59–65

Hobson, J., Wright, J.L. & Churg, A. (1990) Active oxygen species mediate asbestos fiber uptake by tracheal cells. *FASEB J.*, 4, 3135–3139

Huang, S.L. (1979) Amosite, chrysotile and crocidolite asbestos are mutagenic in chinese hamster lung cells. *Mutat. Res.*, 68, 265–274

Huang, S.L., Saggioro, D., Michelmann, H. & Malling, H.V. (1978) Genetic effects of crocidolite asbestos in Chinese hamster lung cells. *Mutat. Res.*, 57, 225–232

Janssen, Y.M.W., Heintz, N.H., Marsh, J.P., Borm, P.J.A. & Mossman, B.T. (1994a) Induction of c-fos and c-jun protooncogenes in target cells of the lung and pleura by carcinogenic fibers. *Am. J. Respir. Cell Mol. Biol.*, 11, 522–530

Janssen, Y.M.W., Marsh, J.P., Absher, M.P., Gabrielson, E., Borm, P.J.A., Driscoll, K. & Mossman, B.T. (1994b) Oxidant stress responses in human pleural mesothelial cells exposed to asbestos. *Am. J. Respir. Crit. Care Med.*, 149, 795–802

Jaurand, M.-C. (1989) Particulate-state carcinogenesis. A survey of recent studies on the mechanism of action of fibres. In: Bignon, J., Peto, J. & Saracci, R., eds, *Non-Occupational Exposure to Mineral Fibres* (IARC Scientific Publications No. 90), Lyon, IARC, pp. 54–73

Jaurand, M.-C. (1991) Mechanisms of fibre genotoxicity. In: *Mechanisms in Fibre Carcinogenesis* (NATO Advanced Workshop), pp. 287–307

Jaurand, M.-C., Bastie-Sigeac, I., Bignon, J. & Stoebner, P. (1983) Effect of chrysotile and crocidolite on the morphology and growth of rat pleural mesothelial cells. *Environ. Res.*, 30, 255–269

Jaurand, M.-C., Kheuang, L., Magne, L. & Bignon, J. (1986) Chromosomal changes induced by chrysotile fibres or benzo(3-4)pyrene in rat pleural mesothelial cells. *Mutat. Res.*, 169, 141–148

Jaurand, M.-C., Fleury, J., Monchaux, G., Nebut, M. & Bignon, J. (1987) Pleural carcinogenic potency of mineral fibers (asbestos, attapulgite) and their cytotoxicity on cultured cells. *J. Natl Cancer Inst.*, 79, 797–804

Kamp, D.W., Graceffa, P., Pryor, W.A. & Weitzman, S.A. (1992) The role of free radicals in asbestos-induced diseases. *Free Rad. Biol. Med.*, 12, 293–315

Kasai, H. & Nishimura, S. (1984) DNA damage induced by asbestos in the presence of hydrogen peroxide. *Gann Monogr. Cancer Res.*, 75, 841–844

Kelsey, K.T., Yano, E., Liber, H.L. & Little, J.B. (1986) The in vitro genetic effects of fibrous erionite and crocidolite asbestos. *Br. J. Cancer*, 54, 107–114

Kenne, K., Ljungquist, S. & Ringertz, N.R. (1986) Effects of asbestos fibers on cell division, cell survival, and formation of thioguanine-resistant mutants in Chinese hamster ovary cells. *Environ. Res.*, 39, 448–464

Kinnula, V.L., Aalto, K., Raivio, K.O., Walles, S. & Linnainmaa, K. (1994) Cytotoxicity of oxidants and asbestos fibers in cultured human mesothelial cells. *Free Rad. Biol. Med.*, 16, 169–176

Kodama, Y., Boreiko, C.J., Maness, S.C. & Hesterberg, T.W. (1993) Cytotoxic and cytogenetic effects of asbestos on human bronchial epithelial cells in culture. *Carcinogenesis*, 14, 691–697

Korkina, L.G., Durnev, A.D., Suslova, T.B., Cheremisina, Z.P., Daugel-Dauge, N.O. & Afanas'ev, I.B. (1992) Oxygen radical-mediated mutagenic effect of asbestos on human lymphocytes suppression by oxygen radical scavengers. *Mutat. Res.*, 265, 245–253

Lasky, J.A., Coin, P.G., Lindroos, P.M., Ostrowski, L.E., Brody, A.R. & Bonner, J.C. (1995) Chrysotile asbestos stimulates platelet-derived growth factor-AA production by rat lung fibroblasts in vitro: evidence for an autocrine loop. *Am. J. Respir. Cell Mol. Biol.*, 12, 162–170

Lavappa, K.S., Fu, M.M. & Epstein, S.S. (1975) Cytogenetic studies on chrysotile asbestos. *Environ. Res.*, 10, 165–173

Leanderson, P., Söderkvist, P., Tagesson, C. & Axelson, O. (1988) Formation of 8-hydroxydeoxyguanosine by asbestos and man made mineral fibers. *Br. J. Ind. Med.*, 45, 309–311

Leanderson, P., Söderkvist, P. & Tagesson, C. (1989) Hydroxyl radical mediated DNA base modification by manmade mineral fibres. *Br. J. Ind. Med.*, 46, 435–438

Lechner, J.F., Tokiwa, T., LaVeck, M., Benedict, W.F., Banks-Schlegel, S., Yeager, H., Barnerjee, A. & Harris, C.C. (1985) Asbestos-associated chromosomal changes in human mesothelial cells. *Proc. Natl Acad. Sci. USA*, 82, 3884–3888

Lemaire, I., Gingras, D. & Lemaire, S. (1982) Thymidine incorporation by lung fibroblasts as a sensitive assay for biological activity of asbestos. *Environ. Res.*, 28, 399–409

Libbus, B.L., Illenye, S.A. & Craighead, J.E. (1989) Induction of DNA strand breaks in cultured rat embryo cells by crocidolite asbestos as assessed by nick translation. *Cancer Res.*, 49, 5713–5718

Linnainmaa, K., Pelin-Enlund, K., Jantunen, K., Vanhala, E., Tuomi, T., Fitzgerald, J. & Yamasaki, H. (1991) Chromosomal damage and GAP junctional intercellular communication in mesothelioma cell lines and cultured human primary mesothelial cells treated with MMMF, asbestos and erionite. In: Brown R.C., Hoskins, J.A. & Johnson, N.F., eds, *Mechanisms in Fibre Carcinogenesis*, New York, Plenum Press, pp. 223, 327–334

Lu, J., Keane, M. J., Ong, T. & Wallace, W.E. (1994) In vitro genotoxicity studies of chrysotile asbestos fibers dispersed in simulated pulmonary surfactant. *Mutat. Res.*, 320, 253–259

Lu, Y.P., Lasne, C., Lowy, R. & Chouroulinkov, I. (1988) Use of the orthogonal design method to study the synergistic effects of asbestos fibres and 12-*O*-tetradecanoylphorbol-13-acetate (TPA) in the BALB/3T3 cell ransformation system. *Mutagenesis*, 3, 355–362

Marsh, J.P. & Mossman, B.T. (1991) Role of asbestos and active oxygen species in activation and expression of ornithine decarboxylase in hamster tracheal epithelial cells. *Cancer Res.*, 51, 167–173

Mast, R.W., McConnell, E.E., Hesterberg, T.W., Chevalier, J., Kotin, P., Thevenaz, P., Bernstein, D.M., Glass, L.R., Miller, W.C. & Anderson, R. (1995a) Multiple-dose chronic inhalation toxicity study of size-seperated kaolin refractory ceramic fiber in male fischer 344 rats. *Inhalation Toxicol.*, 7, 469–502

Mast, R.W., McConnell, E.E., Anderson, R. Chevalier, J., Kotin, P., Bernstein, D.M., Glass, L.R., Miiler, W.C. & Hesterberg, T.W. (1995b) Studies on the chronic toxicity (inhalation) of four types of refractory ceramic fiber in male fischer 344 rats. *Inhalation Toxicol.*, 7, 425–467

McClellan, R.O., Miller, F.J., Hesterberg, T.W., Warheit, D.B., Bunn, W.B., Kane, A.B., Lippmann M., Mast, R.W., McConnell, E.E. & Reinhardt, C.F. (1992) Approaches to evaluating the toxicity and carcinogenicity of man-made fibers: summary of a workshop held November 11–13, 1991, Durham, North Carolina. *Regul. Toxicol. Pharmacol.*, 16, 321–364

McConnell, E.E., (1994) Synthetic vitreous fibers – Inhalation studies. *Regul. Toxicol. Pharmacol.*, 20, S22–S34

McGavran, P.D. & Brody, A.R. (1989) Chrysotile inhalation induces tritiated thymidine incorporation by epithelial cells of distal bronchioles. *Am. J. Respir. Cell Mol. Biol.*, 1, 231–235

Mikalsen, S.O., Rivedal, E. & Sanner, T. (1988) Morphological transformation of Syrian hamster embryo cells induced by mineral fibres and the alleged enhancement of benzo[*a*]pyrene. *Carcinogenesis*, 9, 891–899

Moalli, P.A., Macdonald, J.L., Goodlick, L.A. & Kanr, A.B. (1987) Acute injury and regeneration of the mesothelium in response to asbestos fibers. *Am. J. Pathol.*, 128, 426–445

Mossman, B.T. & Landesman, J.M. (1983) Importance of oxygen free rdicals in asbestos-induced injury to airway epithelial cells. *Chest*, 83, S50–S51

Mossman, B.T., Marsh, J.P. & Shatos, M.A. (1986) Alteration of superoxide dismutase activity in tracheal epithelial cells by asbestos and inhibition of cytotoxicity by antioxidants. *Lab. Invest.*, 54, 204–212

Moyer, V.D., Cistulli, C.A., Vaslet, C.A. & Kane, A.B. (1994) Oxygen radicals and asbestos carcinogenesis. *Environ. Health Perspectives*, 102, 131–136

Olofsson, K. & Mark, J. (1989) Specificity of asbestos-induced chromosomal aberrations in short-term cultures human mesothelial cells. *Cancer Genet. Cytogenet.*, 41, 33–39

Oshimura, M., Hesterberg, T. M., Tsutsui, T. & Barrett, J.C. (1984) Correlation of asbestos-induced cytogenetic effects with cell transformation of Syrian embryo cells in culture. *Cancer Res.*, 44, 5017–5022

Palekar, L.D., Eyre, J.F., Most, B.M. & Coffin, D.L. (1987) Metaphase and anaphase analysis of V79 cells exposed to erionite, UICC chrysotile and UICC crocidolite. *Carcinogenesis*, 8, 553–560

Patérour, M.J., Bignon, J. & Jaurand, M.-C. (1985) In vitro transformation of rat pleural mesothelial cells by chrysotile and/or benzo[a]pyrene. *Carcinogenesis*, 6, 523–529

Pelin, K., Husgafvel-Pursiainen, K., Vallas, M., Vanhala, E. & Linnainmaa, K. (1992) Cytotoxicity and anaphase aberrations induced by mineral fibres in cultured human mesothelial cells. *Toxicol. In Vitro*, 6, 445–450

Pelin, K., Hirvonen, A. & Linnainmaa, K. (1994) Expression of cell adhesion molecules and connexins in gap junctional intercellular communication deficient human mesothelioma tumour cell lines and communication competent primary mesothelial cells. *Carcinogenesis*, 15, 2673–2675

Pelin, K., Hirvonen, A., Taavitsainen, M. & Linanainmaa, K. (1995a) Cytogenetic response to asbestos fibers in cultured human primary mesothelial cells from 10 different donors. *Mutat. Res.*, 334, 225–233

Pelin, K., Kivipensas, P. & Linnainmaa, K. (1995b) Effects of asbestos and man-made vitreous fibers on cell division in cultured human mesothelial cells in comparison to rodent cells. *Environ. Mol. Mutag.*, 25, 118–125

Poole, A., Brown, R.C., Turver, C.J., Skidmore, J.W. & Griffiths, D.M. (1983) In vitro genotoxic activities of fibrous erionite. *Br. J. Cancer*, 47, 697–705

Price-Jones, M.J., Gubbings, G. & Chamberlain, M. (1980) The genetic effects of crocidolite asbestos; comparison of chromosome abnormalities and sister chromatid exchanges. *Mutat. Res.*, 79, 331–336

Reiss, B., Solomon, S., Tong, C., Levenstein, M., Rosenberg, S.H. & Williams, G.M. (1982) Absence of mutagenic activity of three forms of asbestos in liver epithelial cells. *Environ. Res.*, 27, 389–397

Renier, A., Levy, F., Pilliere, F. & Jaurand, M.-C. (1990) Unscheduled DNA synthesis in rat pleural mesothelial cells treated with mineral fibres or benzo[a]pyrene. *Mutat. Res.*, 241, 361–367

Saint-Etienne, L., Endo Capron, S. & Jaurand, M.-C. (1993) In vitro neoplastic transformation of rat pleural mesothelial cells. *Eur. Respir. Rev.*, 3, 141–144

Sincock, A.M., Delhanty, J.D.A. & Casey, G. (1982) A comparison of the cytogenetic response to asbestos and glass fibre in Chinese hamster and human cell lines – demonstration of growth inhibition in primary human fibroblasts. *Mutat. Res.*, 101, 257–268

Sincock, A. & Seabright, M. (1975) Induction of chromosome changes in Chinese hamster cells by exposure to asbestos fibres. *Nature*, 257, 56–58

Smith, D.M., Ortiz, L.W., Archuleta, R.F. & Johnson, N.F. (1987) Long-term health effects in hamsters and rats exposed chronically to man-made vitreous fibres. *Ann. Occup. Hyg.*, 31, 731–754

Suzuki, K. & Hei, T. K. (1996) Induction of heme oxygenase (HO) protein in mammalian cells by mineral fibers: distinctive effect of reactive oxygen species on the induction of HO and cytotoxicity. *Carcinogenesis*, 17, 661–667

Takeuchi, T. & Morimoto, K. (1994) Crocidolite asbestos increased 8-hydroxydeoxyguanosine levels in cellular DNA of a human promyelocytic leukemia cell line, HL60. *Carcinogenesis*, 15, 635–639

Turver, C.J. & Brown, R.C. (1987) The role of catalytic iron in asbestos induced lipid peroxidation and DNA strand breakage in C3H10T1/2 cells. *Br. J. Cancer*, 56, 133–136

Valerio, F., de Ferrari, M., Ottaggio, L., Repetto, E. & Santi, L. (1983) Chromosomal aberrations induced by chrysotile and crocidolite in human lymphocytes in vitro. *Mutat. Res.*, 122, 397–402

Verschaeve, L. & Palmer, P. (1985) On the uptake and genotoxicity of UICC chrysotile A in human primary lung fibroblasts. *Naturwissenschaften*, 72, 326–327

Walker, C., Everitt, J. & Barrett, J.C. (1992) Possible cellular and molecular mechanisms for asbestos carcinogenicity. *Am. J. Ind. Med.*, 21, 253–273

Woodworth, C.D., Mossman, B.T. & Craithead, J.E. (1983) Induction of squamous metaplasia in organ cultures or hamster trachea by naturally occurring and synthetic fibers. *Cancer Res.*, 43, 4906–4912

Wydler, M., Maier, P. & Zbinden, G. (1988) Differential cytotoxic, growth-inhibiting and lipid-peroxidative activities of four different asbestos fibres in vitro. *Toxicol. In Vitro*, 2, 297–302

Yegles, M., Saint-Etienne, L., Renier, A., Janson, X. & Jaurand, M.-C. (1993) Induction of metaphase and anaphase/telophase abnormalities by asbestos fibers in rat pleural mesothelial cells in vitro. *Am. J. Respir. Cell Mol. Biol.*, 9, 186–191

Yegles, M., Janson, X., Dong, H.Y., Renier, A. & Jaurand, M.-C. (1995) Role of fibre characteristics on cytotoxicity and induction of anaphase/telophase aberrations in rat pleural mesothelial cells in vitro. Correlations with in vivo animal findings. *Carcinogenesis*, 16, 2751–2758

Corresponding author
M.-C. Jaurand
INSERM Unité 139, Institut Mondor de Médecine Moléculaire (IM3), Faculté de Médecine, 8 rue du Général Sarrail, 94010 Créteil Cedex, France

Effects of fibres on cell proliferation, cell activation and gene expression

K.E. Driscoll

This document reviews selected aspects of the data bank on fibre toxicology in the respiratory tract and associated tissues. The focus is on processes that may contribute to the neoplastic effects of various fibres in humans and/or animals. However, since non-neoplastic responses, such as inflammation and fibrosis, may contribute to fibre carcinogenesis, studies on these processes are also discussed. The topics reviewed include the following: (i) fibre-induced cell proliferation *in vivo* and the mechanisms underlying proliferative responses; (ii) mechanisms of fibre-induced cell activation, including potential fibre–receptor interactions, signal transduction pathways and oxidative stress; (iii) fibre-induced activation of antioxidant mechanisms; and (iv) proto-oncogene and tumour suppressor gene expression in fibre-induced lung tumours.

Cell proliferation

Exposure of humans and/or animals to some types of fibres has been shown to result in pulmonary fibrosis, lung carcinoma and mesothelioma; diseases that involve, at some stage in their pathogenesis, the proliferation of cells. Neoplastic transformation of cells requires mutation in critical target genes, for example, proto-oncogenes (e.g. c-*myc*, c-*sis*, c-*ras*) and tumour suppressor genes (e.g. *p53*, *Rb1*, *WT1*). Also required for tumour development is the clonal expansion of cells containing critical mutations (Barrett, 1994). In this respect, chemical or physical agents may increase the risk of cancer by multiple mechanisms, including genotoxic effects that result in mutation in cancer genes; or by enhancing cell proliferation, which may increase the frequency of spontaneous mutation and/or promote the clonal expansion of genetically altered cells. The net result of increasing cell proliferation is an increase in the cell populations at risk for the genetic changes associated with neoplastic transformation. In recent years, there has been a renewed interest in the importance of cell proliferation in the carcinogenic process, particularly with respect to how cell proliferation may influence the dose–response relationships between chemical exposure and carcinogenic response.

Understanding the mechanism(s) by which a carcinogenic agent is acting to produce cancer is an important part of the evaluation of potential health risks associated with that exposure (Cohen & Ellwein, 1991). For example, the relationship between exposure and carcinogenesis may differ for (i) chemicals that act to alter expression/function of critical target genes directly (e.g. via mutagenic or clastogenic effects); (ii) chemicals that act by stimulating cell proliferation; or (iii) chemicals that can do both. Also, for materials that act by promoting proliferation of cells, the mechanisms by which this effect is elicited can also be of importance; for example, the nature of the dose–response relationship for carcinogenic responses may differ depending on whether proliferation is the result of a specific receptor-mediated phenomenon or occurs as a compensatory response to tissue injury (Cohen & Ellwein, 1991).

Fibre exposure and cell proliferation: in-vivo studies
Morphological evaluation of animal and/or human tissues has clearly demonstrated hyperplasia of lung epithelium, mesenchymal cells and macrophages, as well as mesothelial cells, after exposure to various fibres. More recent studies using quantitative morphometry or stereology, have provided detailed information on the changes in cell number and tissue volumes in response to asbestos fibres (Chang *et al.*, 1988). During the past 10 years, cell proliferation in response to asbestos and other fibres has been investigated more directly by assessing the incorporation of tritiated deoxythymidine (^{3}HdThd) or 5-bromo-2´deoxyuridine (BrdU) into cells. These studies, which are summarized in Table 1, provide information on the nature and temporal pattern of lung and mesothelial cell

Table 1. Studies characterizing the effect of in-vivo fibre exposure on mesothelial cell, epithelial cell and fibroblast proliferation

Fibre type	Animal species	Treatment	Results	Reference
Mesothelial cell responses				
UICC crocidolite	Mouse	200 µg, intraperitoneal	Short fibres (c. 91% < 2 µm) were cleared by lymphatics and resulted in minimal inflammation and mesothelial cell proliferation (thymidine incorporation). Long fibres (c. 60% > 2 µm) not readily cleared by lymphatics and persisted to 6 months after treatment. Long fibres resulted in accumulation of inflammatory cells, mesothelial cell injury and proliferation.	Moalli et al., 1987
UICC crocidolite	Mouse	3–42 weekly intraperitoneal injections of 200 µg	Inflammation and submesothelial fibrosis associated with fibre clusters after 6–12 weekly injections. Mesothelial cell proliferation (thymidine incorporation) increased progressively over 42 weekly injections. Mesothelial cell proliferation was dissociated spatially and temporally from the development of submesothelial fibrosis.	MacDonald & Kane, 1993
RCF-1, MMVF-10 glass fibre	Rat Hamster	Intrapleural injection: RCF-1, 0.7–0.78×10^6 fibres; MMVF-10, 0.28–14.7×10^6 fibres Intrapleural injection: RCF-1, 0.5–0.56×10^6 fibres; MMVF-10, 0.2–10.5×10^6 fibres	At 7 days and 28 days after exposure, rats and hamsters exhibited similar pleural inflammatory responses, with no remarkable fibre-type dependent differences. Proliferation (BrdU staining) of surface mesothelial cells was greater for hamsters than rats, and was greater in both species for RCF-1 than MMVF-10. Proliferation of mesothelial cells embedded in areas of pleural fibrosis was observed in hamsters but not rats.	Rutten et al., 1994 Everitt et al., 1994
Chrysotile	Mouse	Inhalation, 4 mg/m$^3 \times 5$ h	Proliferation (thymidine incorporation) of mesothelial cells as well as subpleural cells and mesenchymal cells was observed 2 days and 8 days after exposure.	Coin et al., 1991
UICC crocidolite	Mouse	100 µg instilled intratracheally	Short fibres (98% < 2.5 µm) elicited minimal increases in mesothelial cell proliferation (thymidine incorporation). Short fibres did not produce fibrosis. Long fibres (88% > 2.5 µm) produced fibrosis and stimulated proliferation of mesothelial cells that persisted to 2 weeks after exposure. No fibres were found associated with the pleura, suggesting that the effect of fibre exposure on mesothelial cell proliferation was indirect.	Adamson et al., 1993

Table 1 contd. Studies characterizing the effect of in-vivo fibre exposure on mesothelial cell, epithelial cell and fibroblast proliferation

Fibre type	Animal species	Treatment	Results	Reference
UICC crocidolite	Mouse	Intratracheal instillation: 100 µg crocidolite, 1 mg quartz, 3 mg bleomycin; intravenously: 3 mg bleomycin; 90% $O_2 \times 5$ days	Increased proliferation (thymidine incorporation) of mesothelial cells and subpleural cells occurred in response to lung injury by all pneumotoxins. The proliferative response was transient after hyperoxia and persistent after bleomycin, silica and asbestos; the latter three also caused fibrosis.	Adamson et al., 1994
RCF-1	Rat, hamster	6 h/day × 5 days to 6900 fibres & particles/mL (1700 fibres ≥ 5 µm long/mL)	Immediately after 5 days exposure pulmonary neutrophilic inflammation was present in both rats and hamsters. Pleural neutrophilic inflammation was observed in hamsters only. One week after exposure, mesothelial proliferation was detected in rats and hamsters. Mesothelial proliferation (BrdU staining) associated with the visceral pleura was greater for hamsters than for rats.	Everitt et al., 1994.
NIEHS chrysotile	Rat	6h/day, 5 days/week to 0.18mg/m^3 or 8.2 mg/m^3 for 20 days	Proliferation (BrdU staining) of bronchial epithelial cells, interstitial cells and mesothelial cells observed after 5 days of exposure. No significant increase in cell proliferation was detected after 20 days exposure as well as after the 20 days exposure plus 20 days recovery.	Quinlan et al., 1995
Epithelial cell and fibroblast responses				
UICC crocidolite	Mouse	100 µg intratracheal instillation	Acute airway and alveolar inflammation resolved by 2 weeks after exposure. Bronchiolar epithelial cell necrosis at sites of fibre depositin was followed by proliferation (thymidine incorporation) by bronchiolar epithelial cells; the airway proliferative response resolved by 2 weeks. Alveolar type II epithelial cell proliferation was present but less pronounced than airway cells and resolved by 2 weeks. Interstitial cell proliferation was primarily peribronchial and consisted of macrophages (1–2 weeks) and fibroblasts (2–6 weeks).	Adamson & Bowden, 1986
Chrysotile Carbonyl iron	Rat	Chrysotile, 10 mg/m$^3 \times 5$ h; iron, 30 mg/m$^3 \times 5$ h	Chrysotile increased proliferation (thymidine incorporation) of clara cells in the terminal bronchioles and alveolar type II epithelial cells located at the first alveolar duct bifurcation. Epithelial cell proliferation was not detected after	Brody & Overby, 1989; Chang et al., 1988

Table 1 contd. Studies characterizing the effect of in-vivo fibre exposure on mesothelial cell, epithelial cell and fibroblast proliferation

Fibre type	Animal species	Treatment	Results	Reference
			chrysotile exposure of mice and rats. The response resolved by 72 h and returned to control levels by 2 days after exposure. An influx of macrophages to the terminal bronchioles/alveolar ducts occurred coincident with epithelial proliferation. Carbonyl iron exposure did not stimulate epithelial cell proliferation.	
Chrysotile Carbonyl iron	Mouse Rat	Mice: chrysotile, 4 mg/m^3 × 5 h Rats: chrysotile, 10 mg/m^3 × 5 h; carbonyl iron, 30 mg/m^3 × 5 h	Increased proliferation (thymidine incorporation) by endothelial and smooth muscle cells of vessels near the first alveolar duct bifurcation occurred 24 h after chrysotile exposure of mice and rats. The response resolved by 72 h after exposure. Carbonyl iron (rats) did not affect endothelial or smooth muscle cell proliferation. Increased numbers of smooth muscle cells, but no endothelial cells were observed.	McGavran et al., 1990
Chrysotile	Mouse strains: B10.D2/nSn (C5+); B10.D2/oSn (C5−)	4 mg/m^3 × 5 h	Chrysotile exposure resulted in increased alveolar and interstitial macrophages although this response was less in the C5− mice. Proliferation of bronchiolar and alveolar epithelial cells was increased 2–4 days after chrysotile and was similar in both mouse strains. Interstitial fibrosis was observed 1 month after chrysotile exposure in the C5+ but not the C5− mice.	McGavran et al., 1990
NIEHS crocidolite	Rat	7–10 mg/m^3, 6h/day, 5 days/week for 10 and 20 days	Proliferation (thymidine incorporation) occurred in alveolar epithelial cells (fourfold increase) and interstitial cells (threefold increase) after 20 days exposure. Cell proliferation was not restricted to areas of fibrosis. Proliferation of bronchial epithelial cells was not detected.	Mossman et al., 1991
NIEHS chrysotile	Rat	6h/day, 5 days/week to 0.18 mg/m^3 or 8.2 mg/m^3 for 20 days	Proliferation (BrdU staining) of bronchial epithelial cells, interstitial cells and mesothelial cells observed after 5 days of exposure. No significant increase in cell proliferation detected after the 20-day exposure period as well as after the 20-day exposure plus 20-day recovery period.	Quinlan et al., 1995

UICC, Union Internationale Contre le Cancer; RCF, refractory ceramic fibres; MMVF, man-made vitreous fibres; BrdU, bromodeoxyuridine; NIEHS, X; BrdU, 5-bromo-2′deoxyuridine; C5+, C5 sufficient; C5−, C5 deficient.

proliferation in animals after acute or subchronic fibre exposure, as well as insights into the factors that influence fibre-induced cell proliferation.

Acute intratracheal instillation or inhalation of asbestos fibres (i.e. crocidolite or chrysotile) has been shown to result in transient increases in airway and alveolar type II cell proliferation in mice and rats (Adamson & Bowden, 1986; Brody & Overby, 1989; McGavran et al., 1990; Adamson et al., 1993). After a single inhalation exposure, proliferating epithelial cells have been observed at sites of fibre deposition, which includes clara cells in terminal bronchioles and type II alveolar epithelial cells at the first alveolar duct bifurcation. This initiation of epithelial proliferation appears to be coincident with an influx of inflammatory cells. In addition to epithelial cells, interstitial cells (fibroblast-like cells and macrophages), smooth muscle cells and endothelial cells located in the peribronchiolar regions adjacent to sites of epithelial cell necrosis/proliferation and inflammatory cell accumulation proliferate in response to acute asbestos exposure (Adamson & Bowden, 1986; Brody & Overby, 1989; McGavran et al., 1990). Like the epithelial cells, the proliferative response of smooth muscle cells, endothelial cells and interstitial cells is transient, lasting days to a few weeks after an acute inhalation/instillation exposure to high concentrations/doses of fibres.

Mesothelial cell proliferation has been examined in animals after intrapleural or intraperitoneal injection of fibres as well as after respiratory tract exposure by inhalation or intratracheal instillation. Studies by A.B. Kane and co-workers (Moalli et al., 1987) using intraperitoneal injection demonstrated fibre size-dependent proliferative responses of the mesothelium to crocidolite. Specifically, acute intraperitoneal injections of short (91% $\leq$ 2 µm) crocidolite fibres and non-fibrous particulates (i.e. silica, titanium dioxide) were more readily cleared via the lymphatics than injections of long (60% $\geq$ 2 µm) fibres. The short fibres and non-fibrous particles did not provoke remarkable tissue injury or inflammation and caused minimal or no increase in mesothelial cell proliferation. In contrast, long crocidolite fibres accumulated at entry ports of the lymphatics and were associated with marked inflammation, injury to mesothelial cells and mesothelial cell proliferation, the latter response persisting for two weeks after exposure.

Subsequent studies examining the effects of repeated weekly intraperitoneal injections (up to 12 weekly exposures) of a preparation of mixed long and short crocidolite fibres demonstrated a mesothelial cell proliferative response that increased in severity upon repeated exposure (MacDonald & Kane, 1993). Mesothelial proliferation after repeated crocidolite injection appeared dissociated spatially and temporally from pleural fibrosis, suggesting that it was occurring independently of the fibrotic process.

Intrapleural injection studies have identified species differences in the mesothelial response to fibres (Everitt et al., 1994; Rutten et al., 1994). Specifically, intrapleural injection of a refractory ceramic fibre (RCF-1) and a man-made vitreous fibre (MMVF-10) into hamsters and rats was shown to elicit dose-related increases in mesothelial cell proliferation, with the response being greater in hamsters (Everitt et al., 1994; Rutten et al., 1994). In addition, while proliferation of surface mesothelial cells was observed in both rats and hamsters, only in hamsters was proliferation detected in subsurface aggregates of mesothelial cells present in areas of pleural fibrosis. These differences in the magnitude and nature of mesothelial proliferative responses were suggested as a potential factor underlying the markedly greater mesothelioma response of hamsters versus rats to RCF-1 inhalation. In addition to species differences, this intrapleural injection study also demonstrated fibre-specific differences in response, in that the overall proliferative response to RCF-1 was greater than that to MMVF-10. This observation is of particular note since the fibre preparations used were matched for the numbers of long thin fibres administered, suggesting that the greater proliferative response to RCF-1 was due to factors other than the numbers of long fibres (such as surface chemistry of the fibre, or differing amounts of non-fibrous particulate) (Everitt et al., 1994; Rutten et al., 1994).

Recent inhalation and intratracheal instillation studies indicate that secondary mechanisms (i.e., those that do not require a direct interaction between the fibre and the cell) are important in fibre-induced mesothelial cell proliferation, at least after acute high-dose/concentration exposures. For example, Adamson et al. (1993) intratracheally instilled mice with 100 µg of long (88% > 2.5 µm) and short (98% < 2.5 µm) crocidolite and examined lung and mesothelial cell proliferation. Short fibres evoked a minimal inflammatory response, no fibrotic reaction and produced only a slight increase in the proliferation of bronchial epithelial cells, interstitial cells, mesothelial cells and subpleural cells; this proliferation had subsided five days after exposure. In contrast, long fibres produced epithelial cell necrosis and proliferation as well as pulmonary fibrosis and resulted in a pronounced proliferative response of mesothelial and subpleural cells that persisted for two weeks after exposure. Notably, the authors observed no fibres associated with the pleura or subpleural tissues, suggesting an indirect mechanism for the fibre-induced mesothelial cell proliferation (it should be noted that it is not clear with what sensitivity fibres could be detected in the pleura). In a subsequent study, exposure of mice to quartz, bleomycin, crocidolite and hyperoxia were all shown to produce pulmonary inflammation and stimulate the proliferation of mesothelial cells and subpleural fibroblasts (Adamson et al., 1994). The persistence of the proliferative response in this study was greatest for those treatments (i.e. quartz, bleomycin, crocidolite) resulting in pulmonary fibrosis. These observations clearly indicate that lung tissue injury is a stimulus for mesothelial cell proliferation, most likely through the local production of inflammatory lipids and proteins. The findings of Adamson et al. (1993, 1994) are supported by recent observations from short-term inhalation studies with chrysotile in mice and RCF-1 in rats and hamsters which demonstrated increased mesothelial cell proliferation one day to two weeks after fibre exposure (Coin et al., 1991; Everitt et al., 1994). The short time period between fibre inhalation and mesothelial cell proliferation suggests that, like the intratracheal instillation studies, proliferative responses may have occurred in the absence of a direct action of fibres on the mesothelium. Importantly, the above studies focused on responses after single intratracheal exposures or up to five daily inhalation exposures. More recently, Quinlan et al. (1995) reported on the proliferative effects of chrysotile inhalation for 6 h per day, five days a week for up to three weeks. In the three-week inhalation study, increased proliferation of epithelial, interstitial and mesothelial cells was detected after five days of chrysotile exposure; however, with continued fibre inhalation, the proliferative response decreased to a point where it was not different from the air control animals, suggesting adaptation to continued fibre exposure.

The studies described above provide insights into factors that influence fibre-induced cell proliferation and raise questions on the relationship between cell proliferation and fibre-induced carcinogenesis, as well as the utility of cell proliferation data for assessing the potential toxicity of fibres. Regarding the latter, it is important to note that, with the exception of the work by MacDonald and Kane (1993), the studies reported to date have been of short duration. In this respect, it will be important to characterize further the nature and temporal pattern of cell proliferative responses after more prolonged fibre exposures. The question of short- versus long-term effects is underscored by the observation that lung cell proliferation returned toward control levels with continuing chrysotile exposure (Quinlan et al., 1995). In addition to the limited information on effects after longer-term exposure, another weakness with the current data bank is that many of the studies employed nonphysiological routes of exposure and/or high bolus doses, and only a few fibre types have been examined. Regarding the use of nonphysiological exposure methods and high-dose levels, these types of studies can provide insights into mechanisms of action; however, it is important that any findings made can be shown to be reasonably relevant to real-life exposure situations. Regarding the cell proliferation data, it is difficult to use the existing data bank to predict the effects of chronic low-level fibre exposure or even compare broadly across various classes of fibres.

Recognizing that there are gaps in the existing data bank, studies conducted to date have clearly demonstrated that exposure to fibres can stimulate the proliferation of a variety of lung cell populations. In addition, and consistent with previous observations on fibre toxicity and carcinogenicity, it appears to be the long crocidolite fibres that have the greatest effect on both lung and mesothelial cell proliferation. Regarding the role of proliferation in fibre-induced carcinogenesis, the above studies raise some interesting issues by demonstrating the following: (i) multiple cell types, including those not associated with fibre-induced neoplasia, proliferate in response to fibre exposure; (ii) mesothelial cell proliferation appears to occur as a general response to lung injury and is not specific to fibres; and (iii) respiratory tract exposure to RCF-1 fibres elicits significant mesothelial cell proliferation in both hamsters and rats, although only the former develop a marked mesothelioma response to RCF-1. Overall, these observations appear to dissociate fibre-induced cell proliferation *per se* from the development of carcinoma and mesothelioma and point to the importance of additional factors in fibre-induced cancer. In addition, these studies raise questions as to whether cell proliferation data (at least from short-term exposure studies) provide useful information for assessing potential fibre carcinogenicity. In this respect, since one key reason for characterizing cell proliferation derives from its conceptual role in promoting genotoxic effects, an additional and perhaps more meaningful approach for evaluating the carcinogenic potential of fibres may be to characterize relationships between in-vivo fibre exposure and genotoxicity in key target cell populations. The availability of transgenic models for evaluating mutagenesis *in vivo* (Gorelick, 1995) and the development of methods for characterizing mutation in the *hprt* gene of rat lung cells after in-vivo chemical exposure (Driscoll *et al.*, 1995a, 1996) provide approaches to characterize more directly relationships between *in vivo* fibre exposure and mutagenic effects in lung cells.

Qualitative and quantitative differences were observed between the hamster and rat in the proliferative response of mesothelial cells to RCF-1 and MMVF-10 exposure. In addition, in response to inhaled RCF-1, hamsters, but not rats, developed pleural inflammation (Everitt *et al.*, 1994). However, it is not clear whether the magnitude and nature of these differences can explain the dramatic species differences in mesothelioma response to RCF-1 (i.e. 43% incidence of mesothelioma in RCF-1-exposed hamsters and 3% incidence in rats). The marked species difference in mesothelioma response and the apparently limited differences in proliferative response point again to the importance of factors, in addition to cell proliferation, in the carcinogenic effects of some fibres. In this respect, species comparisons after longer-term exposures may reveal greater differences in the magnitude and nature of proliferation. Alternatively, species differences in fibre clearance and translocation, DNA repair, cytokine/growth factor production or sensitivity of target cells to these mediators (as suggested by the species difference in pleural inflammation seen after RCF-1) may contribute to differences in tumorigenic response and will be important to characterize. In this respect, a key issue regarding species susceptibility is which test animal species, if any, represents an appropriate model for assessing fibre associated health hazards to humans.

Mechanisms of fibre-induced cell proliferation
There are likely to be multiple and overlapping mechanisms underlying the cell proliferative response to fibre exposure with the precise mechanism(s) dominating a particular response influenced by: the magnitude, duration and method of exposure; the type and characteristics of the fibres involved; and the tissues affected. In general, key mechanisms thought to underlie fibre-induced cell proliferation include the following: a wound-healing response secondary to direct fibre-induced cell injury; fibre-induced activation of inflammatory and other lung cells by noncytotoxic mechanisms to release mediators that cause tissue injury and/or stimulate cell proliferation; and a direct mitogenic effect of fibres on target cells.

Tissue injury initiates a wound-healing response involving a cascade of events resulting in the recruitment and activation of inflammatory cells, cell proliferation, extracellular matrix synthesis and, if the injurious agent is not persistent and/or exposure is not recurrent, the resolution of the response. The inflammatory, proliferative and biosynthetic aspects of the wound-healing response are orchestrated by a complex network of bioactive lipids and proteins, the latter including cytokines and growth regulatory factors (Rom *et al.*, 1991; Driscoll, 1995). That the proliferative response to fibres occurs, under some conditions, as a response to direct fibre-induced tissue injury can be inferred from the known cytotoxic activity of some fibres *in vitro* (e.g. chrysotile) and observations that initial sites of fibre-induced cell proliferation, particularly in acute high-dose studies, correspond to sites of epithelial necrosis (Adamson *et al.*, 1993, 1994). At present, the properties of fibres responsible for cytotoxic activity and, thus, direct tissue injury are incompletely understood;

however, it is clear that both physical and chemical factors are important (Harrington et al., 1975; Goodlick & Kane, 1986; Marsh & Mossman, 1988; Driscoll, 1994). For example, long chrysotile fibres are more cytotoxic than short fibres (Marsh & Mossman, 1988) and early studies indicated that the magnesium content (and probably positive surface charge) of chrysotile contributed to its membranolytic effects (Harrington et al., 1975). In addition, the iron content of fibres such as crocidolite and amosite (which can be up to 36% iron by weight) can contribute to cytotoxicity by catalysing production of toxic oxygen species such as hydroxyl radicals (OH$^{\bullet}$) (Goodlick & Kane,

Table 2. Effect of in-vitro fibre exposure on the production of bioactive lipids and proteins

Target cell	Fibre type	Mediator	Possible role of the cytokine in fibre-induced toxicity	References
Macrophage	Crocidolite, amosite, chrysotile, RCF-1	Tumour necrosis factor α	Recruitment of inflammatory cells; activate production of chemokines (e.g. IL-8, MIP-2); stimulate growth factor (e.g. PDGF) production.	Driscoll et al., 1990; Perkins et al., 1991; Zhang et al., 1993; Leikauf et al., 1994
Macrophage	Crocidolite	Interleukin-1	Recruitment of inflammatory cells; stimulate lymphocyte proliferation; activate production of chemokines (e.g. IL-8, MIP-2); stimulate growth factor (e.g. PDGF) production.	Oghiso & Kubota, 1986; Zhang et al., 1993
Macrophage	Crocidolite, chrysotile	Leukotriene B4	Recruitment of inflammatory cells	Dubois et al., 1989; Driscoll et al., 1990
Macrophage	Chrysotile	Insulin growth factor-1 PDGF	Stimulate proliferation of fibroblasts and mesothelial cells.	Noble et al., 1991
Macrophage	Chrysotile	PDGF	Stimulate proliferation of fibroblasts, mesothelial cells, endothelial cells and smooth muscle cells. Chemotaxis of fibroblasts and smooth muscle cells.	Bonner et al., 1991
Fibroblasts	Chrysotile	PDGF	Stimulate proliferation of fibroblasts, mesothelial cells, endothelial cells and smooth muscle cells. Chemotaxis of fibroblasts and smooth muscle cells.	Bonner et al., 1993; Lasky et al., 1995
Alveolar epithelial cells	Crocidolite, chrysotile	Interleukin-8	Recruitment of inflammatory cells; stimulate endothelial cell proliferation (angiogenesis).	Rosenthal et al., 1994
Alveolar epithelial cells	Crocidolite (MMVF-10 glass fibre was not active)	MIP-2	Recruitment of inflammatory cells; stimulate proliferation of epithelial cells.	Driscoll et al., 1995c
Pleural mesothelial cells	Amosite, chrysotile, crocidolite	Interleukin-8	Recruitment of inflammatory cells; stimulate endothelial cell proliferation (angiogenesis).	Boylan et al., 1992; Griffith et al., 1994
Pleural mesothelial cells	Amosite, chrysotile, crocidolite	Interleukin-1	Recruitment of inflammatory cells; activate production of chemokines (e.g. IL-8, MIP-2); stimulate growth factor (e.g. PDGF) production.	Griffith et al., 1994

RCF, refractory ceramic fibre; IL-8, interleukin-8; MIP-2, macrophage inflammatory protein-2; PDGF, platelet-derived growth factor.

1986; Shatos et al., 1987). Further, the ability of some fibres (e.g. crocidolite, erionite) to accumulate iron from endogenous sources, making it available for redox cycling and the generation of reactive oxygen species (ROS) can be a factor in cytotoxicity and genotoxicity (Chao et al, 1994; Eborn & Aust, 1995).

There is considerable evidence that some fibres cause inflammatory and proliferative effects by directly activating production of pro-inflammatory and growth regulatory mediators within the lung and pleura (Bauman et al., 1990; Driscoll et al., 1990; Donaldson et al., 1992; Rosenthal et al., 1994). For example, the inflammatory response to chrysotile asbestos results, at least in part, from its ability to activate a complimentary pathway that generates the anaphylotoxin C5a, a potent chemotactic and activating factor for inflammatory cells (Warheit et al., 1985). In addition, exposure of macrophages to subcytotoxic doses of several fibre types (e.g. crocidolite, chrysotile, erionite, glass fibre 100 (GF 100), sepiolite) activates production of ROS that can injure cells, produce genotoxic effects (Donaldson & Cullen, 1984; Hansen & Mossman, 1987) or, as discussed below, stimulate cell activation (Janssen et al., 1995a; Driscoll et al., 1995b). As summarized in Table 2, there is an increasing number of studies demonstrating that noncytotoxic concentrations of various fibres interact with lung macrophages, epithelial cells, fibroblasts and mesothelial cells to stimulate the production of pro-inflammatory and growth regulatory mediators. In this respect, the recruitment and activation of inflammatory cells and the local release of ROS, inflammatory lipids and cytokines is recognized as a critical component in the pathogenesis of many chronic interstitial lung diseases, including those occurring in response to asbestos and other fibres (for a review see Rom et al., 1991). The importance of cell activation in the pathogenesis of fibre-induced and other interstitial diseases is supported by observations from animal and clinical studies demonstrating the presence of inflammatory cells in the lung or pleura that are releasing pro-inflammatory and growth regulatory factors, including many of those listed in Table 2 (Garcia et al., 1989; Rom, 1991; Driscoll, 1995). At present, there is only limited information on the characteristics of fibres that contribute to their ability to activate macrophages and other lung cells. Fibrous shape has been shown to be important for some responses in that mineral fibres (i.e. crocidolite, erionite, glass fibre) but not their non-fibrous analogues activate macrophage production of ROS (Hansen & Mossman, 1987). Since cell activation appears to occur at exposure levels that do not produce cytotoxicity, it is likely that activation responses (as opposed to overt cytotoxicity) are most relevant to fibre-induced effects at the low levels of exposure. In this respect, there is a need to define further those chemical and physical factors that influence fibre-induced cell activation.

The degree to which proliferative responses to fibre exposure arise as a compensatory response to overt cytotoxicity or as a consequence of fibre-induced cell activation will be influenced by a number of factors, most notably the nature of the exposure. Clearly high fibre doses and exposure rates (for example, as occurs with bolus intratracheal instillation, intraperitoneal and intrapleural dosing, or acute inhalation of high fibre concentrations) are more likely to overwhelm lung defences and elicit direct cytotoxic effects. This is an important issue since, as documented for low-solubility non-fibrous particles, even materials that are relatively innocuous produce adverse lung responses characterized by inflammatory cell recruitment, cytokine production, cell proliferation, fibrosis and tumours when administered at high enough doses (Oberdörster, 1994). Importantly, the nature of the response and the profile of mediators produced after exposure to high doses of low-toxicity materials or lower doses of inherently toxic materials can be similar, making it difficult to discern different mechanisms of action in high-dose studies (Driscoll & Maurer, 1991). This raises issues regarding the extrapolation of responses/mechanisms from high to low fibre exposures. In this respect, one reason that the fibre-induced cell proliferation responses characterized to date do not appear to correlate with carcinogenicity may be that, under the exposure conditions used, any fibre-specific mechanisms are masked by a generic response to high doses (or dose rates) of material.

An additional mechanism by which fibres may stimulate cell proliferation is through a direct mitogenic action on target cells. Table 3 summarizes several studies demonstrating that various fibres stimulate cell proliferation when added to cells or

Table 3. In-vitro effects of fibres on cell proliferation

Test materials	Target cell	Treatment	Key findings	Reference
UICC crocidolite	Pleural explants (mesothelial)	0.01% suspension of fibres	Fibre exposure of human pleural explants resulted in marked proliferation of mesothelial cells.	Rajan et al., 1972
UICC amosite, crocidolite	Hamster tracheal explant	4 mg/mL in 60 mm dishes ± retinyl methyl ether	Asbestos fibres produced metaplasia and proliferation of epithelial cells. Response was attenuated by retinoids.	Mossman et al., 1980
UICC crocidolite, chrysotile, haematite	Hamster tracheal epithelial cells	Crocidolite and haematite: 0.06–0.13 µg/cm^2; 0.6–1.3 µg/cm^2	Crocidolite and chrysotile, but not haematite, stimulated ODC activity; asbestos fibres stimulated cell proliferation.	Landesman & Mossman, 1982
UICC crocidolite, chrysotile, GF 100, riebeckite, antigorite, glass beads	Hamster tracheal epithelial cells	0.3–2.6 µg/cm^2	Crocidolite, chrysotile and glass fibres all stimulated ODC activity. Non-fibrous analogues did not stimulate ODC. Long chrysotile fibres were more active than short fibres. Fibre-activated ODC was inhibited by calcium antagonists and PKC inhibitors.	Marsh & Mossman, 1988
UICC crocidolite, chrysotile, GF 100, riebeckite, antigorite, glass beads	Hamster tracheal epithelial cells	Crocidolite, 0.01 µg/cm^2; riebeckite, 20 µg/cm^2; chrysotile, 0.01–1 µg/cm^2	In-vitro toxicity: chrysotile > glass fibre > crocidolite > riebeckite > antigorite > glass beads. Increased colony-forming efficiency: crocidolite > antigorite ~ riebeckite > chrysotile, glass fibre, glass beads. TGFβ decreased the colony-forming efficiency response to crocidolite.	Sesko & Mossman, 1989
UICC crocidolite	Hamster tracheal explants	4 mg/mL	Crocidolite produced squamous metaplasia in explant tissues. Effect inhibited by DFMO (inhibitor of ODC) and retinal acetate.	Cameron et al., 1989
UICC crocidolite, riebeckite glass beads	Hamster tracheal epithelial cells	0.1–1.0 µg/cm^2	Mitogenic concentrations of crocidolite increase diacylglycerol suggesting a mechanism for fibre-activated PKC. Glass beads and riebeckite did not increase diacylglycerol or PKC.	Sesko et al. 1990
Crocidolite, chrysotile, para-aramid	Hamster tracheal epithelial cells, rat lung fibroblasts	0.025–5.0 µg/cm^2	Crocidolite, chrysotile and to a lesser extent aramid fibres stimulated proliferation (thymidine incorporation, colony forming efficiency) and increased ODC activity of epithelial cells but not fibroblasts.	Marsh et al., 1994

Table 3 contd. In-vitro effects of fibres on cell proliferation

Test materials	Target cell	Treatment	Key findings	Reference
NIEHS crocidolite	Hamster tracheal epithelial cells	0.25, 2.5 and 5.0 µg/cm^2	Crocidolite and H_2O_2 increased expression of a c-*jun*-promoter luciferase construct. Over-expression of c-*jun* produced by transfection with a c-*jun* expression construct resulted in increased proliferation and anchorage dependent growth of hamster tracheal epithelial cells.	Timblin *et al.*, 1995
NIEHS chrysotile	Rat lung fibroblast	0.01–10 µg/cm^2	Chrysotile increased proliferation of quiescent fibroblasts. The response was attenuated by addition of anti-platelet-derived growth factor antibody.	Lasky *et al.*, 1995

UICC, Union Internationale Contre le Cancer; ODC, ornithine decarboxylase; GF, glass fibre; PKC, protein kinase C; TGFβ, transforming growth factor β; DFMO, difluoromethylornithine; NIEHS, US National Institute of Environmental Health Sciences.

tissue explants *in vitro*. The cell types shown to exhibit proliferative responses to fibres include human mesothelial cells, hamster tracheal epithelial cells and rat lung fibroblasts. Studies on mechanism of fibre-induced mitogenesis indicate that this response may involve the following: the activation of intracellular signalling pathways that mimic those of growth factors (Sesko *et al.*, 1990, Heintz *et al.*, 1993; Timblin *et al.*, 1995); the stimulation of growth factor production and autocrine growth regulation (Lasky *et al.*, 1995); and the up-regulation of growth factor receptor expression (Bonner *et al.*, 1993).

Exposure of hamster tracheal epithelial cells to crocidolite, chrysotile and glass fibre (but not their non-fibrous analogues) activates the enzyme ornithine decarboxylase, which is rate-limiting in the synthesis of polyamines – growth regulatory molecules necessary for cell proliferation and differentiation (Marsh & Mossman, 1988; Pegg, 1988). Induction of ornithine decarboxylase by crocidolite is dependent, at least in part, on calcium and activation of protein kinase C (Marsh & Mossman, 1988). Crocidolite and chrysotile also induce expression of the proto-oncogenes c-*fos* and c-*jun* in mesothelial cells but only c-*jun* in tracheal epithelial cells (Heintz *et al.*, 1993). Regarding the latter, persistent overexpression of c-*jun* in tracheal epithelial cells has been shown to stimulate their proliferation (Timblin *et al.*, 1995). With respect to the proliferation of fibroblasts, Lasky *et al.* (1995) have shown that in-vitro chrysotile exposure induces the expression of platelet-derived growth factor (PDGF)-A chain mRNA and the release of PDGF protein by the fibroblasts, which act *in vitro* as autocrine regulators of proliferation. In addition, in-vitro exposure of fibroblasts to chrysotile increases the expression of the PDGF-A receptor (PDGF RA) causing these cells to become more responsive to the mitogenic effects of PDGF. The effect on PDGF receptors is not unique to fibres – iron beads also increase PDGF RA expression (Bonner *et al.*, 1993). In a manner similar to fibre-induced cell activation, the direct effects of fibres on the proliferation of epithelial cells and fibroblasts has been reported to occur at noncytotoxic exposure levels. Thus, this mechanism of fibre-induced cell proliferation may also be important at low levels of in-vivo fibre exposure.

Fibre-induced cell activation

As discussed above, some fibres directly activate cells *in vitro* to proliferate and/or release cytotoxic or pro-inflammatory mediators. Given that fibre-induced cell activation has been reported to occur at exposure levels below those eliciting cytotoxic effects, it is likely that activation responses play a critical role in fibre-induced lung effects, particularly at low fibre burdens. In this respect, understanding of the mechanisms of fibre-induced cell activation should provide insights into the chemical and physical properties of fibres that contribute to their bioactivity in environmental exposure situations. This type of information will be useful to guide the development and interpretation of mechanistically based screening assays designed to assess the toxic potential of fibres. At present, the precise mechanisms by which fibres activate cells are incompletely understood; however, in-vitro studies conducted primarily with asbestos provide insights into the nature of fibre–cell interactions, some of the intracellular signalling pathways involved in the activation process and the contribution of oxidants to activation responses. An important question to be answered as this mechanistic data bank is expanded to include non-asbestos fibres is whether there exist common activation pathways among fibres leading to a given response (i.e. growth factor release, oxidant production, proliferation) or whether there are multiple fibre-specific cell activation mechanisms. The extent to which common pathways exist will determine to a large extent the utility of screening assays for these pathways in assessing potential fibre toxicity.

Interaction of fibres with cell membranes
In-vitro studies on asbestos have identified some mechanisms by which fibres interact with cell membranes resulting in cell activation. Positively charged chrysotile fibres bind to negatively charged sialic acid residues on erythrocyte and macrophage cell membranes, and this interaction mediates, at least in part, the lysis of the erythrocytes and the activation of eicosanoid biosynthesis by macrophages (Kouzan *et al.*, 1985; Gallagher *et al.*, 1987). Binding to sialic acid and the subsequent activation of macrophages, however, is not unique to chrysotile; positively charged non-fibrous particles (i.e. iron beads) show similar effects (Kouzan *et al.*, 1985). Given that sialic acid can be associated with proteins on many cell types, binding of positively charged particles via this mechanism may be relevant to a variety of cells. More recently, crocidolite fibres were shown to bind to type I and type II macrophage scavenger receptors (Resnick *et al.*, 1993). Type I and II receptors bind a wide range of polyanionic ligands; that the interaction of crocidolite with macrophage scavenger receptors is associated with its negative surface charge was demonstrated by the ability of various anionic ligands to block fibre binding (Resnick *et al.*, 1993). Regarding cell activation, it is noteworthy that binding to scavenger receptors *per se* does not appear to activate macrophages (Kobzik *et al.*, 1993). Fibre surface charge can also facilitate membrane receptor binding by promoting fibre opsonization. For example, opsonization of chrysotile and crocidolite fibres with IgG immunoglobulin is influenced by fibre surface charge and results in the binding of the immunoglobulin-coated fibres to macrophage Fc receptors, which activates cells (Scheule & Holian, 1989, 1990; Perkins *et al.*, 1991). Opsonization-mediated receptor binding is not unique to fibres – non-fibrous particles (e.g. silica, aluminium beads) can act in a similar manner (Perkins *et al.*, 1991). Overall, these studies indicate that surface charge is a factor in facilitating interactions between asbestos fibres (as well as other charged particles) and the cell membrane.

The studies described above indicate that asbestos can interact with cell membrane molecules/receptors directly or indirectly (i.e. secondary to fibre opsonization). The extent to which similar mechanisms of fibre–cell interaction exist for other fibre types is not known; however, since binding to sialic acid and macrophage scavenger receptors, and opsonization by IgG, also occur with non-fibrous particles, these mechanisms probably apply to a wide range of positively or negatively charged materials. Considering interactions with sialic acid residues or scavenger receptors, it is not clear whether particle binding via these mechanisms (i) stimulates a specific receptor-mediated signalling event or (ii) serves primarily to bring the fibre into more intimate contact with the cell where the fibre's physical and/or chemical properties, and in some manner (for example, the catalysis of oxidant production) perturb the cell membrane and cause activation. It should be noted that studies relating fibre–cell membrane interactions to cell activation have focused largely on macrophages and, specifically, eicosanoid metabolism and oxidant production by these cells. Additional work is therefore needed to determine whether similar mechanisms are relevant to other cell types (e.g. mesothelial cells, epithelial cells) as well as other fibre-induced responses (e.g. proliferation, cytokine release).

Signal transduction pathways in fibre-induced cell activation

As summarized in Table 4, multiple signalling pathways appear to be activated by asbestos exposure. Studies by Mossman and co-workers (Marsh & Mossman, 1988; Sesko et al., 1990; Perderiset et al., 1991) have demonstrated that in-vitro exposure of hamster tracheal epithelial cells to asbestos activates protein kinase C, a response associated with increased phosphatidyl inositol turnover and generation of diacylglycerol. Inhibiting various components of this pathway, as well as the influx of extracellular calcium, blocks asbestos-induced increases in epithelial ornithine decarboxylase activity (Marsh & Mossman, 1988; Sesko et al., 1990). Like epithelial cells, asbestos-induced activation (i.e. O_2^- production) of alveolar macrophages is dependent, at least in part, on increased phosphatidyl inositol turnover, increases in cytosolic calcium (derived from both intra- and extracellular sources) and activation of protein kinase C (Roney & Holian, 1989; Kalla et al., 1990). More recently, crocidolite-induced activation of c-*fos* and c-*jun* expression in mesothelial cells has been shown to involve protein kinase C as well as mitogen-activated protein kinase (Fung et al., 1995; Zanella et al., 1995). Thus, a signalling pathway involving protein kinase C appears to be important in various cell types exposed to asbestos. The demonstration that the protein kinase C response is associated with increases in diacylglycerol indicates that an early event in asbestos-induced cell activation is the activation of phospholipase C. Also, the observation that asbestos increases cytosolic calcium and that calcium channel blockers attenuate some asbestos-induced cell activation responses (i.e. increased ornithine decarboxylase, oxidant production) implies a role for calcium-mediated signalling.

To date, research on signalling mechanisms has focused primarily on asbestos and, therefore, it is uncertain whether other fibres act in a similar manner. However, recent studies indicate that both similarities and differences exist between asbestos and non-asbestos fibres in their ability to activate cells. For example, exposure of epithelial cells to biogenic silica fibres increases ornithine decarboxylase activity and this response can be inhibited by calcium channel blockers. In addition, recent studies demonstrate that crocidolite and chrysotile induce a delayed but persistent expression of c-*jun* and c-*fos* in pleural mesothelial cells and a delayed and persistent induction of c-*jun*, but not c-*fos*, in hamster tracheal epithelial cells (Heintz et al., 1993). Subsequent studies have demonstrated that the non-asbestos fibres, MMVF-10 glass fibre and RCF-1, as well as non-fibrous riebeckite particles, do not induce expression of c-*jun* or c-*fos* in hamster tracheal epithelial cells at exposure levels comparable (in mass) to those effective for asbestos (Janssen et al., 1994b). In contrast to epithelial cells, mesothelial cells did respond to MMVF-10 and RCF-1, as well as erionite fibres, with increased c-*fos* and c-*jun* m-RNA (Janssen et al., 1994a). However, the dose (in mass) of MMVF-10 and RCF-1 required to elicit the mesothelial proto-oncogene response was 10-fold greater than that needed for crocidolite, chrysotile or erionite. These studies indicate that cell type-specific differences exist in fibre-induced expression of immediate early genes (i.e. c-*jun* in epithelial cells versus c-*jun* and c-*fos* in mesothelial cells). In addition, there appear to be differences in potency with which fibres activate c-*jun* and c-*fos* expression. The proto-oncogenes c-*fos* and c-*jun* code for proteins that can form homo- and heterodimeric complexes comprising the activator protein 1 (AP-1) transcription factors. AP-1 binds to specific sites in DNA and can stimulate transcription of a wide range of genes including several that are required for cell division. Interpreting the significance of the fibre-induced c-*fos* and/or c-*jun* response will require a better understanding of the consequences of increases in these proto-oncogenes in fibre-stimulated cells. In this respect, very recent studies have shown that increases in c-*jun* expression in tracheal epithelial cells stimulates cell proliferation (Timblin et al., 1995).

Oxidative stress and fibre-induced cell activation

There is increasing evidence that oxygen radical formation catalysed directly or indirectly by asbestos fibres is a key factor underlying cytotoxic and cell activation responses. One mechanism by which asbestos fibres can generate ROS is through redox reactions catalysed by metals on the surface of the fibres (Lund & Aust, 1991). Studies in cell-free systems demonstrate that asbestos gives rise to OH• when incubated in the presence of hydrogen

Table 4. Effect of in-vitro fibre exposure on intracellular signalling mechanisms

Signalling event	Cell type	Fibre type	Key observations	Reference
Calcium mobilization	Hamster tracheal epithelial cell	Chrysotile	Treatment of cells with calcium channel blockers (e.g. verapamil) attenuated asbestos-induced ODC activity.	Marsh & Mossman, 1988
Calcium mobilization	Guinea-pig alveolar macrophage	Chrysotile	Asbestos increased cytosolic calcium levels. Treatment with calcium channel blocker (i.e. verapamil) attenuated asbestos-induced increases in cytosolic calcium and oxidant production. Increasing extracellular calcium levels enhanced asbestos-induced oxidant production.	Roney & Holian, 1989; Kalla et al., 1990
Calcium mobilization	Mouse epidermis	Biogenic silica fibres	Silica fibres increased ODC activity. The effect was inhibited by calcium channel blockers.	Bhatt et al., 1994
Phosphatidyl inositol turnover	Guinea-pig alveolar macrophage	Chrysotile	Asbestos increased phosphatidyl inositol metabolism, generating inositol triphosphate and diacylglycerol.	Roney & Holian, 1989
Phosphatidyl inositol turnover	Hamster tracheal epithelial cell	Crocidolite, riebeckite, glass beads	Asbestos increased phosphatidyl inositol metabolism generating inositol triphosphate and diacylglycerol. Riebeckite and glass beads did not stimulate this effect.	Sesko et al., 1990
PKC activation	Hamster tracheal epithelial cell	Chrysotile	Treatment of cells with PKC inhibitors (H-7, H-8) attenuated asbestos-induced ODC.	Marsh & Mossman, 1988
PKC activation	Guinea-pig alveolar macrophage	Chrysotile	Asbestos increased PKC activity.	Roney & Holian, 1989
PKC activation	Hamster tracheal epithelial cell	Crocidolite	Asbestos increased PKC activity.	Perderiset et al., 1991
PKC activation	Rat pleural mesothelial cells	Crocidolite	Inhibition of PKC (calphostin C) attenuated asbestos-induced expression of c-fos and c-jun.	Fung et al., 1995
Mitogen activated kinase	Rat pleural mesothelial cells	Crocidolite, chrysotile	Asbestos stimulated phosphorylation of MAP kinase.	Zanella et al., 1995
Induction of c-fos and c-jun expression; increased AP-1 activity	Rat pleural mesothelial cells; hamster tracheal epithelial cells	Crocidolite, chrysotile, polystyrene beads	Crocidolite and chrysotile, but not polystyrene beads, elicited a persistent induction of c-fos and c-jun mRNA in mesothelial cells. Asbestos induced a persistent expression of c-jun in tracheal epithelial cells, but had no detectable effect on c-fos. Asbestos increased levels of AP-1 activity in both cell types.	Heintz et al., 1993

Table 4 contd. Effect of in-vitro fibre exposure on intracellular signalling mechanisms

Signalling event	Cell type	Fibre type	Key observations	Reference
Induction of c-fos and c-jun expression	Rat pleural mesothelial cells; hamster tracheal epithelial cells	Crocidolite, chrysotile, MMVF-10 glass fibre; RCF-1; erionite; riebeckite	Crocidolite and chrysotile induced expression of c-jun in epithelial cells; at comparable mass doses MMVF-10, RCF-1 and riebeckite did not. Crocidolite, erionite and chrysotile increased expression of c-fos and c-jun in mesothelial cells; at 10-fold higher concentrations, MMVF-10 and RCF-1 increased c-fos and c-jun expression.	Janssen et al., 1994b
Increased NFκB activity	Hamster tracheal epithelial cells	Crocidolite	Asbestos increased nuclear levels of NFκB binding activity in tracheal epithelial cells. Pretreatment with N-acetylcysteine attenuated this effect.	Janssen et al., 1995a
Increased NFκB activity	Rat alveolar epithelial cells	Crocidolite, MMVF-10 glass fibre, quartz, titanium dioxide	Crocidolite and quartz increased nuclear NFκB binding activity in alveolar epithelial cells. Glass fibre and titanium dioxide did not.	Driscoll et al., 1995c

ODC, ornithine decarboxylase; PKC, protein kinase C; MAP, macrophage activating protein; AP, adenosine phosphate; MMVF, man-made vitreous fibre; RCF, refractory ceramic fibre; NFκB, nuclear factor κB.

peroxide (Weitzman & Graceffa, 1984) and produces lipid peroxidation when added to aqueous solutions of phospholipid (Weitzman & Weitberg, 1985). Treatment of the asbestos with the iron chelator desferrioxamine eliminates the oxidant-generating activity of the fibres, indicating that iron plays a critical role in direct ROS production. In this respect, in-vitro mobilization of iron from crocidolite via the action of low molecular weight chelators greatly increases the associated generation of ROS and, thus, iron mobilization may be an important factor in fibre-mediated oxidative stress in vivo (Lund & Aust, 1992). Further, recent studies indicate that erionite and crocidolite can acquire iron and make it available for redox cycling and generation of ROS (Eborn & Aust, 1995). This finding suggests a mechanism whereby particulates initially containing a minimal amount of iron or no iron at all may facilitate iron-catalysed production of ROS. In addition to direct fibre-mediated effects, production of ROS can occur as a result of fibres interacting with phagocytic cells and stimulating an oxidative burst. As reviewed by Kamp et al. (1992), macrophages and neutrophils from a variety of species respond to asbestos as well as other fibres (i.e. GF 100, erionite) with the production of superoxide anion and hydrogen peroxide. Fibre exposure may also increase oxidant generation by stimulating cytokine production. For example, asbestos is known to activate macrophages and epithelial cells to release pro-inflammatory cytokines such as tumour necrosis factor α (TNFα) and interleukin-8 (IL-8) which can stimulate an oxidative burst by phagocytic cells.

There is compelling evidence that ROS play a key role in asbestos-induced cytotoxicity. Briefly, the addition of antioxidant enzymes (e.g. superoxide dismutase (SOD), catalase) or oxidant-scavenging molecules (i.e., mannitol, dimethylthiourea) to cells in culture has been shown to attenuate the cytotoxic effects of asbestos on tracheal epithelial cells (Mossman et al. 1986), macrophages (Goodlick &

Kane, 1986), fibroblasts (Shatos et al., 1987) and endothelial cells (Garcia et al., 1989). Treatment of the asbestos with the iron chelator desferrioxamine or addition of desferrioxamine to cell cultures also inhibits asbestos-induced toxicity in many in-vitro systems, indicating that iron-catalysed oxidant production is an important factor (Goodlick & Kane, 1986; Shatos et al., 1987). Interestingly, in tracheal epithelial cell cultures, the protective effects of antioxidants were observed for long chrysotile or crocidolite fibres but not for glass fibres or short (≤ 2 µm) chrysotile fibres (Shatos et al., 1987); this suggests that factors in addition to oxidants are important in the toxicity of some fibres or fibre preparations.

In addition to cytotoxicity, the ability of some fibres to catalyse the production of ROS is important in cell activation. For example, crocidolite-induced activation of ornithine decarboxylase in tracheal epithelial cells can be blocked by catalase (Marsh & Mossman, 1991). In these same studies, addition of hydrogen peroxide or the oxidant generating enzyme system, xanthine/xanthine oxidase, also increased ornithine decarboxylase, further demonstrating the importance of oxidants in this response. More recent studies indicate that oxidative stress associated with in-vitro asbestos exposure plays a role in activating the expression of pro-inflammatory cytokines by alveolar macrophages and epithelial cells (Rosenthal et al., 1994; Simeonova & Luster, 1995; Driscoll et al., 1995c). In-vitro exposure of rat alveolar macrophages to noncytotoxic concentrations of crocidolite and chrysotile increases TNFα gene expression and protein production (Dubois et al., 1989; Driscoll et al., 1990; Simeonova & Luster, 1995). This effect on TNFα production can be inhibited by OH• scavengers tetramethylthiourea (TMTU) and dimethyl sulfoxide (DMSO) as well as desferrioxamine, the latter implicating iron-catalysed generation of ROS (Simeonova & Luster, 1995). In addition, treatment of macrophages with hydrogen peroxide or systems generating the superoxide anion, hydrogen peroxide and OH• also stimulated macrophage TNFα production. Regarding TNFα, recent studies have shown that high doses of RCF-1 can stimulate the release of this cytokine by macrophages *in vitro*; however, a role for oxidant stress in this response was not examined (Leikauf et al., 1995). Studies using cultures of alveolar epithelial cells have shown that exposure to crocidolite, but not MMVF-10 glass fibre, increases production of macrophage inflammatory protein-2 (MIP-2), a neutrophil chemoattractant and epithelial cell mitogen (Driscoll et al., 1995c, in press). Like the macrophage response to TNFα, crocidolite-induced MIP-2 expression by epithelial cells can be attenuated by OH• scavengers, such as DMSO, ethanol, mannitol and TMTU.

Consistent with a role for oxidative stress in asbestos-induced inflammatory cytokine gene expression are recent studies on the nuclear translocation of nuclear factor κB (NFκB) activation in cells after asbestos exposure. NFκB is a heterodimeric protein complex that regulates transcription of several genes associated with inflammatory and immune responses including TNFα, IL-8 and MIP-2 (Schreck et al., 1991). Numerous studies have demonstrated that activation of NFκB can be affected by oxidative stress. In addition, as shown for asbestos, a variety of antioxidants can attenuate activation of NFκB. In this respect, recent studies demonstrate that crocidolite exposure increases nuclear NFκB levels in hamster tracheal epithelial cells and this response is attenuated by treatment with the sulfhydryl reagent N-acetylcysteine (Janssen et al., 1995a). Crocidolite exposure of alveolar epithelial cells has also been shown to activate NFκB. OH• scavengers (i.e. DMSO, mannitol, ethanol, TMTU) attenuate asbestos-activated NFκB in epithelial cells as well as expression of the NFκB regulated gene MIP-2 (Driscoll et al., 1995c). In addition to NFκB, expression of the proto-oncogenes c-*fos* and c-*jun* can be influenced by the redox status of cells (Abate et al., 1990). As discussed above, exposure of mesothelial cells to crocidolite increases the expression of c-*fos* and c-*jun*. Recent studies indicate that the crocidolite-induced c-*fos* and c-*jun* response occurs in association with decreases in intracellular glutathione, which is indicative of oxidative stress, and can be attenuated by treating the cells with the sulfhydryl reagent, N-acetylcysteine (Janssen et al., 1995b). Interestingly, direct treatment of mesothelial cells with hydrogen peroxide did not decrease cellular glutathione, suggesting that there are differences in oxidant stress produced by insoluble fibres compared to soluble oxidants (Janssen et al., 1995b). Overall, there is growing evidence that asbestos fibres activate cells, at least in part, through oxidative mechanisms. Although the ability of other fibres to activate cells via oxidants has not been examined, there is evidence that some inflammatory non-fibrous particles (i.e. crystalline silica) act through an oxidant-signalling mechanism suggesting that this activation pathway is not unique to asbestos (Driscoll *et al.*, 1995b,d).

Fibre induction of antioxidant defences

The effects of fibre exposure on the expression of enzymes and proteins associated with antioxidant defences has been characterized in several in-vitro and in-vivo studies. One of the first studies in this area demonstrated that in-vitro exposure of hamster tracheal epithelial cells to crocidolite and chrysotile, but not to GF 100, increased SOD activity (Mossman *et al.*, 1986). In-vitro studies on human mesothelial cells and lung fibroblasts indicate that they respond to crocidolite and chrysotile with slight increases in haeme oxygenase; however, the greatest effect on this enzyme, as well as on MnSOD, was produced by exposure to superoxide anion and hydrogen peroxide via xanthine/xanthine oxidase (Janssen *et al.*, 1994b). In-vivo studies have shown that exposure of rats to asbestos increases expression of mRNA for antioxidant enzymes – specifically, MnSOD, CuZnSOD and glutathione peroxidase (Janssen *et al.*, 1992). The MnSOD in crocidolite-exposed rats is primarily in alveolar type II cells and, to a lesser extent, fibroblasts (Holley *et al.*, 1992). In contrast to crocidolite, in-vivo exposure to cristobalite silica induces marked expression of lung MnSOD with minimal effects on other antioxidant enzymes, which suggests particle-specific effects on antioxidant enzymes. In this respect, treatment of tracheal epithelial cells with hydrogen peroxide has been shown to increase catalase primarily, while exposure to the superoxide anion/hydrogen peroxide-generating systems induces MnSOD (Shull *et al.*, 1991); thus, different patterns of particle-induced antioxidant enzymes may reflect different profiles of oxidants or other mediators being released. Interestingly, MnSOD expression in rats after in-vivo exposure to fibrous and non-fibrous particles corresponds to the magnitude and temporal pattern of pulmonary inflammatory response (Janssen *et al.*, 1992, 1994b). This association may reflect the fact both these responses (i.e. MnSOD induction and inflammation) can be elicited by the pro-inflammatory cytokines TNFα and IL-1, which are released by macrophages in response to asbestos and other particulate inflammagens (Dubois *et al.*, 1989; Driscoll *et al.*, 1990; Donaldson *et al.*, 1992; Driscoll, 1995).

The findings on asbestos-induced antioxidant enzyme expression further support a role for oxidants in asbestos toxicity. In this respect, increased expression and activity of various antioxidant enzymes provides a mechanism by which initial exposures to asbestos may increase the lung's ability to cope with subsequent exposures. Supporting this concept is the observation that lung cell proliferation in the rat increases after five days of chrysotile inhalation but returns to control levels with continued exposure (Quinlan *et al.*, 1995). It is noteworthy that in this same study lung antioxidant defences (i.e. MnSOD expression) increased throughout the three-week chrysotile exposure period. Regarding the potential effects of non-asbestos fibres on antioxidant defences, only the effects of glass fibres have been examined to date. GF 100, in contrast to crocidolite and asbestos, did not stimulate SOD activity in tracheal epithelial cells. This observation appears consistent with the finding that the cytotoxic effects of GF 100 on tracheal epithelial cells are not blocked by SOD, mannitol or DMSO (Shatos *et al.*, 1987). It is also consistent with the finding that MMVF-10 glass fibres do not induce expression of the oxidant-inducible cytokine MIP-2 by rat lung epithelial cells (Driscoll *et al.*, in press)

Proto-oncogenes and tumour suppressor genes in asbestos-induced tumours

Primary mesotheliomas and mesothelioma-derived cell lines have been reported to overexpress genes that may contribute to deregulated growth (reviewed by Fitzpatrick *et al.*, 1995). One growth factor, the expression of which has been frequently shown to be altered in human mesothelioma cells, is PDGF. Normal human mesothelial cells express little or no PDGF A or B chain message; however, these PDGF mRNAs, and the corresponding proteins, are overexpressed in several human mesothelioma-derived cell lines (Gerwin *et al.*, 1987; Versnel *et al.*, 1988). Mesothelioma cells have also been shown to express PDGF receptors (Versnel *et al.*, 1991). Murine mesotheliomas, like those in humans, are associated with increased

PDGF expression (Christmas et al., 1993). In this respect, recent studies suggest that PDGF can act as an autocrine growth factor in human and murine mesothelioma cells (Garlepp et al., 1993; Van der Meeren et al., 1993; Dorai et al., 1994). In contrast to human and mouse cells, rat mesothelial cells or mesothelioma cell lines (the latter derived from asbestos-induced and spontaneously occurring mesotheliomas) do not express PDGF mRNA nor do they secrete detectable PDGF protein, even though they express PDGF-B receptors (Walker et al., 1992); these findings suggest that autocrine regulation of mesothelioma cell growth by PDGF is species-specific.

Transforming growth factor (TGF) β is mitogenic for human mesothelial cells and high levels of TGFβ1 and 2 have been observed in pleural effusions from mesothelioma patients (Maeda et al., 1994). However, expression of TGFβ1 mRNA and secretion of active TGFβ protein appear to be similar in normal human mesothelial cells and mesothelioma cells, suggesting that autocrine growth stimulation by TGFβ alone may not be a critical factor in generation of mesothelioma (Gerwin et al., 1987). However, studies using murine mesothelioma cells indicate that reducing TGF production inhibits anchorage-independent growth and retards tumorigenicity, and that the tumours that arise from TGFβ-deficient cells are associated with increased lymphocyte infiltration; the latter observation suggests that a possible role of TGFβ production by murine mesothelioma cells is to suppress immune responses to the developing tumour (Fitzpatrick et al., 1995).

Insulin growth factor I (IGF-I) and TGFβ are two cytokines associated with autocrine growth regulation of mesothelioma cells from humans and/or rodents. Human mesothelial cells respond to IGF-I with increased proliferation, and normal human mesothelial and mesothelioma cells express IGF-I mRNA, secrete IGF-I protein, and express IGF-I receptors (Lee et al., 1991). Studies on immortalized rat mesothelial cells and mesothelioma cell lines have demonstrated IGF expression; however, it appears that the transforming agent influences the development of an IGF-producing phenotype. Specifically, immortalized cell lines derived from normal rat mesothelial cells or from spontaneously arising rat mesotheliomas have been associated with IGF-II gene expression and protein production, and growth of these cells was inhibited by antibody to IGF. In contrast, rat cell lines derived from asbestos-induced mesotheliomas did not express IGF mRNA or produce protein (Rutten et al., 1995). When TGFα expression was examined in these rat cell lines, it was shown that only the lines derived from asbestos-induced mesotheliomas expressed TGFα mRNA and secreted TGFα protein (Walker et al., 1995). In addition, TGFα was shown to inhibit the growth of spontaneously transformed cells but stimulate proliferation of asbestos-transformed mesothelioma cells; the growth of the latter was attenuated by the addition of antibody to TGFα to cell cultures. Overall, these studies suggest that the processes underlying changes in autocrine growth regulation of rat mesothelial cells can be influenced by the transforming agent.

Several additional studies have examined expression of proto-oncogenes in asbestos-transformed cells or lung tumours associated with asbestos exposure. Kanner et al. (1988) reported increased c-src tyrosine kinase activity in asbestos-transformed tumour-derived Syrian hamster embryo (SHE) cells, and the same line was shown to possess an activated Ha-ras gene (Gilmer et al., 1988). However, the increased tyrosine kinase activity and Ha-ras expression were not unique to asbestos transformation, in that similar changes were detected in SHE cells transformed by diethylstilboestrol and benzo[a]pyrene. A study of 20 human mesothelioma cells lines did not detect mutations in K-ras, suggesting that K-ras activation does not constitute a critical step in the development of human mesothelioma.

Altered expression of the tumour suppressor gene p53 has been reported in lung carcinomas from asbestos-exposed individuals (Nuorva et al., 1994). Specifically, increased immunoreactive p53 was associated with 67% of tumours from patients with asbestos exposure versus 40% of carcinoma patients without a history of asbestos exposure. In this study, there was also a positive association between strongly p53-immunostaining tumours and a history of cigarette smoking. Altered expression of p53 has been demonstrated in a murine model of crocidolite-induced mesothelioma (Cora & Kane, 1993). Briefly, an altered p53 gene was detected in 76% of cell lines derived from crocidolite-induced mesotheliomas and decreases in p53 expression were observed in 35% of the neoplastic lines, suggesting that alterations in *p53* contribute to the development of mesotheliomas in this model. Regarding human mesotheliomas, although some studies have reported altered *p53* expression (Cote *et al.*, 1991), this appears to be an infrequent occurrence (Metcalf *et al.*, 1992) suggesting that changes in *p53* are not critical in mesothelioma development in humans.

In considering growth factor/oncogene expression by immortalized and/or tumour-derived cells, it should be noted that the phenotype and/or genotype characterized has developed after considerable in-vivo and/or in-vitro selection. In this respect, the alterations observed may have occurred as early or late events in the neoplastic process and, in the end, the observation of a particular gene being over- or underexpressed may provide limited (if any insight) into the specific mechanisms by which asbestos or some other agent is acting to initiate or promote the neoplastic process. A possible exception to this may be the studies on asbestos-transformed and spontaneously transformed rat mesothelial cells, which indicate differences in autocrine growth regulation dependent on the transforming agent (or lack thereof). However, what these data tell us with respect to early asbestos-mediated events or how specific this differential pattern of growth factor expression is to transformation by fibres as opposed to other carcinogens is unclear at present.

Limitations in the in-vivo and in-vitro approaches

In considering the use of existing information on cell proliferation, cell activation and cell signalling for assessing health hazards and health risks from fibre exposure, a number of limitations in the existing data bank become apparent. Many of these limitations have been discussed in the various sections above, and only a few key issues will be iterated here. Regarding the in-vivo proliferation studies, much if not all of the work to date has been performed using exposure levels and dose rates (for example as occurs in intratracheal, intrapleural or intraperitoneal injection) well above those that would be encountered occupationally or environmentally; thus, the question of dose-dependent mechanisms becomes important. For example, as discussed in the context of cell proliferation and cell activation, several potential mechanisms exist, and the extent that one or all of these mechanisms contribute to a given response will be influenced by the fibre dose and dose-rate. Responses seen at high exposures (i.e. those that greatly exceed human exposure) may be a consequence of overwhelming defence mechanisms that would not be overcome at lower exposures. Therefore, in the context of hazard identification and risk assessment, it becomes important to demonstrate that responses identified under high-exposure conditions are occurring by mechanisms relevant to real-world exposures.

The use of nonphysiological methods of exposure presents another limitation with some of the existing in-vivo data. Studies performed on fibres administered by intratracheal instillation, intraperitoneal injection or intrapleural injection methods can provide insights into mechanisms of action. However, care must be taken when extrapolating observations made using these exposure techniques to potential health hazards under physiological exposure conditions. Clearly, the dose, the condition (e.g. opsonized, unopsonized) and type (i.e. size) of fibres that interact with the target tissue and the handling of the fibre (i.e. redistribution and clearance) can differ depending on the exposure method. The question again becomes whether the mechanisms of action underlying a given response after intratracheal instillation, intrapleural injection or intraperitoneal injection of fibres are relevant to mechanisms occurring under expected human exposure conditions. Other gaps in the in-vivo proliferation data bank include the paucity of longer-term studies as well as the limited data on non-asbestos fibres.

In general, the in-vitro studies have provided and continue to provide useful insights into fibre-induced cell activation and cell signalling mechanisms and have assisted in identifying fibre characteristics contributing to bioactivity (e.g. dimension, iron content, oxidant production). There are, however, a number of technical limitations that frequently preclude a clear extrapolation between in-vitro studies as well as from the in-vitro to the in-vivo situation. For example, the characteristics of fibre preparations tested are frequently not described in sufficient detail to allow comparison between different studies using the same fibre type but different preparations. In addition, static in-vitro cell assays may be insensitive to inherent differences in fibre dissolution or clearance behaviours that are important *in vivo*; the potential issue is that different fibres may exhibit similar effects *in vitro*, but would elicit markedly different responses *in vivo*. Further, a variety of primary cultures or immortalized cell lines are used to evaluate fibre bioactivity. In many instances, these cells require markedly different culture conditions that confound a clear identification of true cell type- or species-specific differences in response versus differences reflecting unique culture conditions. In addition, for most of the in-vitro cell culture models used, there exists limited information on how dose and response reflect potential effects that may occur in intact tissue. Finally, it is essential to confirm mechanisms identified *in vitro* with appropriate in-vivo studies, for example, as is being done in studies on antioxidant defences, cytokine and proto-oncogene expression.

References

Abate, C., Patel, L., Rauscher, F.J. & Curran, T. (1990) Redox regulation of fos and jun DNA-binding activity *in vitro*. *Science*, 249, 1157–1161

Adamson, I.Y.R. & Bowden, D.H. (1986) Crocidolite-induced pulmonary fibrosis in mice cytokinetic and biochemical studies. *Am. J. Pathol.*, 122, 261–267

Adamson, I.Y.R., Bakowska, J. & Bowden, D.H. (1993) Mesothelial cell proliferation after instillation of long or short asbestos fibers into mouse lung. *Am. J. Pathol.*, 142, 1209–1216

Adamson, I.Y.R., Bakowska, J. & Bowden, D.H. (1994) Mesothelial cells proliferation: a nonspecific response to lung injury associated with fibrosis. *Am. J. Pathol.*, 10, 253–258

Barrett, J.C. (1994) Cellular and moleculr mechanisms of asbestos carcinogenicity: implications for biopersistence. *Environ. Health Perspectives*, 102, 19–23

Bauman, M.D., Jetten, A.M., Bonner, J.C., Kumar, R.K., Bennet, R.A. & Brody, A.R. (1990) Secretion of a platelet-derived growth factor homologue by rat alveolar macrophages exposed to particulates in vitro. *Eur. J. Cell Biol.*, 51, 327–334

Bhatt, T.S., Battalora, M. & DiGiovanni, J. (1994) Modulation of biogenic silica fiber-induced ornithine decarboxylase activity. *Proc. Am. Assoc. Cancer Res.*, 35, 172

Bonner, J.C., Osornio-Vargas, A.R., Badgett, A. & Brody, A.R. (1991) Differential proliferation of rat lung fibroblasts induced by the platelet-derived growth factor-AA, -AB, and -BB isoforms secreted by rat alveolar macrophages. *Am. J. Respir. Cell. Mol. Biol.*, 5, 539–547

Bonner, J.C., Goodell, A.L., Coin, P.G. & Brody, A.R. (1993) Chrysotile asbestos upregulates gene expression and production of receptors for platelet-derived growth factor (PDGF-AA) on rat lung fibroblasts. *J. Clin. Invest.*, 92, 425–430

Boylan, A.M., Ruegg, C., Kim, K.J., Hebert, C.A., Hoeffel, J.M., Pytela, R., Sheppard, D., Goldstein, I.M. & Broaddus, V.C. (1992) Evidence of a role for mesothelial cell-derived interleukin-8 in the pathogenesis of asbestos-induced pleurisy in rabbits. *J. Clin. Invest.*, 89, 1257–1267

Brody, A.R. & Overby, L.H. (1989) Incorporation of tritiated thymidine by epithelial and interstitial cells in bronchiolar-alveolar regions of asbestos exposed rats. *Am. J. Pathol.*, 134, 133–140

Cameron, G., Woodworth, C.D., Edmondson, S. & Mosman, B.T. (1989) Mechanisms of asbestos-induced squamous metaplasia in tracheobronchial epithelial cells. *Environ. Health Perspectives*, 80, 101–108

Chang, L.-Y., Overby, L.H., Brody, A.R. & Crapo, J.D. (1988) Progressive lung cell reactions and extracellular matrix production after brief exposure to asbestos. *Am. J. Pathol.*, 131, 156–170

Chao, C.-C., Lund, L.G., Rim, K.R. & Aust, A.E. (1994) Iron mobilization from crocidolite asbestos by human lung carcinoma cells. *Arch. Biochem. Biophys.*, 314, 384–391

Christmas, T.I., Mutsaets, S.E., Manning, L.S., Davis, M.R. & Hoogsteden, B.W.S. (1993) Platelet-derived growth factor as an autocrine factor in murine malignant mesothelioma. *Eur. Respir. Rev.*, 3, 192–194

Cohen, S.M. & Ellwein, L.B. (1991) Genetic error, cell proliferation and carcinogenesis. *Cancer Res.*, 51, 6493–6505

Coin, P.G., Moore, L.B., Roggli, V. & Brody, A.R. (1991) Pleural incorporation of ^{3}H-thymidine after inhalation of chrysotile asbestos in the mouse. *Am. Rev. Respir. Dis.*, 143, A603

Cora, E.M. & Kane, A.B. (1993) Alterations in a tumor suppressor gene, p53, in mouse mesotheliomas induced by crocidolite. *Eur. Respir. Rev.*, 3, 148–150

Cote, R.J., Jhanwar, S.C., Novick, S. & Pelicer, A. (1991) Genetic alterations of the p53 gene are a feature of malignant mesotheliomas. *Cancer Res.*, 51, 5410–5416

Donaldson, K. & Cullen, R.T. (1984) Chemiluminescence of asbestos activated macrophages. *Br. J. Exp. Pathol.*, 65, 81–90

Donaldson, K., Li, X.Y., Dogra, S., Miller, B.G. & Brown, G.M. (1992) Asbestos-stimulated tumour necrosis factor release from alveolar macrophages depends on fibre length and opsonization. *J. Pathol.*, 168, 243–248

Dorai, T., Kobayashi, H., Holland, J.F. & Ohnuma, T. (1994) Modulation of platelet-derived growth factor-B mRNA expression and cell growth in a human mesothelioma cell line by a hammerhead ribozyme. *Mol. Pharmacol.*, 46, 437–444

Driscoll, K.E. (1994) *In vitro* evaluation of mineral cytotoxicity and inflammatory activity. *Rev. Mineral.*, 28, 489–511

Driscoll, K.E. (1995) Role of cytokines in pulmonary inflammation and fibrosis. In: McClellan, R.O. & Henderson, R., eds, *Concepts in Inhalation Toxicology*, Washington DC, Taylor and Francis, pp. 471–504

Driscoll, K.E. & Maurer, J.K. (1991) Cytokine and growth factor release by alveolar macrophages: potential biomarkers of pulmonary toxicity. *Toxicol. Pathol.*, 19, 398–405

Driscoll, K.E., Higgins, J.M., Leytart, M.J. & Crosby, L.L. (1990) Differential effects of mineral dusts on the *in vitro* activation of alveolar macrophage eicosanoid and cytokine release. *Toxicol. In Vitro*, 4, 284–288

Driscoll, K.E., Deyo, L.C., Howard, B.W., Poynter, J. & Carter, J.M. (1995a) Characterizing mutagenesis in the hprt gene of rat alveolar epithelial cells. *Exp. Lung Res.*, 21, 941–956

Driscoll, K.E., Maurer, J.K., Higgins, J. & Poynter, J. (1995b) Alveolar macrophage cytokine and growth factor production in a rat model of crocidolite-induced pulmonary inflammation and fibrosis. *J. Toxicol. Environ. Health*, 46, 155–169

Driscoll, K.E., Howard, B.W., Carter, L.M. & Hassenbein, D.G. (1995c) Inflammatory cytokine production by rat lung epithelial cells *in vitro*: particles, NFκB and oxidative stress. In: *5th International Inhalation Symposium, Hannover Medical School, 1995*

Driscoll, K.E., Hassenbein, D.G., Carter, J.M., Kunkel, S.L., Quinlan, T.R. & Mossman, B.T. (1995d) TNFα and increased chemokine expression in rat lung after particle exposure. *Toxicol. Lett.*, 82/83, 483–489

Driscoll, K.E., Carter, J.M., Howard, B.W., Hassenbein, D., Pepelko, W., Baggs, R. & Oberdorster, G. (1996) Pulmonary inflammatory and mutagenic responses to subchronic inhalation of carbon black by rats. *Toxicol. Appl. Pharm.*, 136, 372–380

Driscoll, K.E., Howard, B.W., Carter, J.M. Asquith, T.A. Detilleux, P., Johnston, C., Kunkel, S.L., Paugh, D. & Isfort, R.I. Chemokine expression by rat lung epithelial cells: effects of *in vitro* and *in vivo* mineral dust exposure. *Am. J. Pathol.* (in press)

Dubois, C.M., Bissonnette, E. & Rola-Pleszczynski, M. (1989) Asbestos fibers and silica particles stimulate rat alveolar macrophages to release tumor necrosis factor. *Am. Rev. Respir. Dis.*, 139, 1257–1264

Eborn, S.K. & Aust, A. (1995) Effect of iron acquisition on induction of DNA single-strand breaks by erionite, a carcinogenic mineral fiber. *Arch. Biochem. Biophys.*, 316, 507–514

Everitt, J.I., Bermudez, E., Mangum, J.B., Wong, B., Moss, O.R., Janszen, D. & Rutten, A.A.J.J.L. (1994) Pleural lesions in Syrian golden hamsters and Fischer-344 rats following intrapleural instillation of man-made ceramic or glass fibers. *Toxicol. Pathol.*, 22, 229–236

Fitzpatrick, D.R., Peroni, D.J. & Bielefelt-Ohmann, H. (1995) The role of growth factors and cytokines in the tumorigenesis and immunobiology of malignant mesothelioma. *Am. J. Respir. Cell. Mol. Biol.*, 12, 455–460

Fung, H., Quinlan, T.R., Janssen, Y.M.W., Timblin, C., Marsh, J.P., Heintzm N., Taatjes, D., Jaken, S. & Mossman, B.T. (1995) Inhibition of protein kinase C (PKC) reduces asbestos-induced c-fos and c-jun expression in rat pleural mesothelial (RPM) cells. *FASEB J.*, 9, A1066

Gallagher, J.E., George, G. & Brody, A.R. (1987) Sialic acid mediates the initial binding of positively charged inorganic particles to alveolar macrophages membranes. *Am. Rev. Respir. Dis.*, 135, 1345–1352

Garcia, J.G.N., Griffith, D.E., Cohen, A.B. & Callahan, K.S. (1989) Alveolar macrophages from patients with asbestos exposure release increased levels of leukotriene B4. *Am. Rev. Respir. Dis.*, 139, 1494–1501

Garlepp, M.J., Christmas, T.I., Mutsaers, S.E., Manning, L.S., Davis, M.R. & Robinson, B.W.S. (1993) Platelet-derived growth factor as an autocrine factor in murine malignant mesothelioma. *Environ. Respir. Rev.*, 3, 192–194

Gerwin, B.I., Lechner, J.F., Reddel, R.R., Roberts, A.B., Robbins, K.C., Gabrielson, E.W. & Harris, C.C. (1987) Comparison of production of transforming growth factor β and platelet-derived growth factor by normal human mesothelial cells and mesothelioma cell lines. *Cancer Res.*, 47, 6180–6184

Gilmer, T.M., Annab, L.A. & Barret, J.C. (1988) Characterization of activated proto-oncogenes in chemically transformed Syrian hamster embryo cells. *Mol. Carcinog.*, 1, 180–188

Goodlick, L.A. & Kane, A.B. (1986) Role of reactive oxygen metabolites in crocidolite asbestos toxicity to mouse macrophages. *Cancer Res.*, 46, 5558–5566

Gorelick, N.J. (1995) Overview of mutation assays in transgenic mice for routine testing. *Environ. Mol. Mutag.*, 25, 218–230

Griffith, D.E., Miller, E.J., Gray, L.D., Idell, S. & Johnson, A.R. (1994) Interleukin-1-mediated release of interluekin-8 by asbestos-stimulated human pleural mesothelial cells. *Am. J. Respir. Cell. Mol. Biol.*, 10, 245–252

Hansen, K. & Mossman, B.T. (1987) Generation of superoxide from alveolar macrophages exposed to asbestiform and nonfibrous particles. *Cancer Res.*, 47, 1681–1686

Harrington, J.S., Allison, A.C. & Badami, X. (1975) Mineral fibers: chemical, physiochemical and biological properties. *Adv. Pharm. Chem.*, 12, 291–402

Heintz, N., Janssen, Y.M.W. & Mossman, B.T. (1993) Persistent induction of c-fos and c-jun protoncogene expression by asbestos. *Proc. Natl Acad. Sci. USA*, 90, 3299–3303

Holley, J.A., Janssen, Y.M.W., Mossman, B.T. & Taajes, T.J. (1992) Increased manganese superoxide dismutase protein in type II epithelial cells of rat lungs after inhalation of crocidolite asbestos or cristobalite silica. *Am. J. Pathol.*, 141, 475–485

Janssen, Y.M.W., Marsh, J.P., Absher, M.P., Hemenway, D., Vacek, P.M., Leslie, K.O., Borm, P.J.A. & Mossman, B.T. (1992) Expression of antioxidant enzymes in rat lungs after inhalation of asbestos or silica. *J. Biol. Chem.*, 267, 10625–10639

Janssen, Y.M.W., Heintz, N.H., Marsh, J.P., Borm, P.J.A. & Mossman, B.T. (1994a) Induction of c-fos and c-jun protooncogenes in target cells of the lung and pleura by carcinogenic fibers. *Am. J. Respir. Cell. Mol. Biol.*, 11, 522–530

Janssen, Y.M.W., Marsh, J.P., Absher, M.P., Gabrielson, E., Borm, P.J.A., Driscoll, K.E. & Mossman, B.T. (1994b) Oxidant stress responses in human pleural mesothelial cells exposed to asbestos. *Am. J. Respir. Clin. Care Med.*, 149, 795–802

Janssen, Y.M.W., Barchowsky, A., Treadwell, M., Driscoll, K.E. & Mossman, B.T. (1995a) Asbestos induces nuclear factor κB (NFκB) DNA-binding activity and NFκB-dependent gene expression in tracheal epithelial cells. *Proc. Natl Acad. Sci. USA*, 92, 8458–8462

Janssen, Y.M.W., Heintz, N.H. & Mossman, B.T. (1995b) Induction of c-fos and c-jun proto-oncogene expression by asbestos is ameliorated by N-acetyl-L-cysteine in mesothelial cells. *Cancer Res.*, 55, 2085–2089

Kalla, B., Hamilton, R.F., Scheule, R.K. & Holian, A. (1990) Role of extracellular calcium in chrysotile asbestos stimulation of alveolar macrophages. *Toxicol. Appl. Pharmacol.*, 104, 130–138

Kamp, D.W., Graceffa, P., Pryor, W.A. & Weitzman, S.A. (1992) The role of free radicals in asbestos-induced diseases. *Free Rad. Biol. Med.*, 12, 293–315

Kanner, S.B., Parsons, S.J., Parsons, J.T. & Gilmer, T.M. (1988) Activation of pp60[c-src] tyrosine kinase specific activity in tumor-derived Syrian hamster embryo cells. *Oncogene*, 2, 327–335

Kobzik, L., Huang, S., Paulauskis, J. & Godleski, J. (1993) Particle opsonization and lung macrophage cytokine response: *in vitro* and *in vivo* analysis. *J. Immunol.*, 151, 2753–2760

Kouzan, S., Gallagher, J.E., Eling, T. & Brody, A.R. (1985) Binding of iron beads to sialic acid residues on macrophage membranes stimulates arachidonic acid metabolism. *Lab. Invest.*, 53, 320–327

Landesman, J.M. & Mossman, B.T. (1982) Induction of ornithine decarboxylase in hamster tracheal epithelial cells exposed to asbestos and 12-O-tetradecanoylphorbol-13-acetate. *Cancer Res.*, 42, 3669–3675

Lasky, J.A., Coin, P.G., Lindroos, P.M., Ostrowski, L.E., Brody, A.R. & Bonner, J.C. (1995) Chrysotile asbestos stimulates platelet-derived growth factor-AA production by rat lung fibroblasts *in vitro*: evidence for an autocrine loop. *Am. J. Respir. Cell. Mol. Biol.*, 12, 162–170

Lee, T.C., Zhang, Y., Aston, C., Hintz, R., Jagirdar, J., Perle, M.A., Burt, M. & Rome, W.N. (1991) Normal human mesothelial cells and mesothelial cell lines express insulin-like growth factor I and associated molecules. *Cancer Res.*, 53, 2858–2864

Leikauf, G.D., Fink, S.P., Miller, M.L., Lockey, J.E. & Driscoll, K.E. (1995) Refractory ceramic fibers activate alveolar macrophage eicosanoid and cytokine release. *J. Appl. Physiol.*, 78, 164–171

Lund, L.G. & Aust, A.E. (1991) Iron-catalyzed reactions may be responsible for the biochemical and biological effects of asbestos. *Biofactors*, 3, 83–89

MacDonald, J.L. & Kane, A.B. (1993) Regulation of mesothelial cell proliferation by the extracellular matrix *in vivo*. *Eur. Respir. Rev.*, 3, 123–125

Maeda, J., Ueki, N., Ohkawa, T., Iwahasi, N., Nakano, T., Hada, T. & Higashino, K. (1994) Transforming growth factor-β1- and -β2-like activity in pleural effusions caused by malignant mesothelioma or primary lung cancer. *Clin. Exp. Immunol.*, 98, 319–322

Marsh, J.P. & Mossman, B.T. (1988) Mechanisms of induction of ornithine decarboxylase activity in tracheal epithelial cells by asbestiform minerals. *Cancer Res.*, 48, 709–714

Marsh, J.P. & Mossman, B.T. (1991) Role of asbestos and active oxygen species in activation and expression of ornithine decarboxylase in hamster tracheal epithelial cells. *Cancer Res.*, 51, 167–173

Marsh, J.P., Mossman, B.T., Driscoll, K.E., Schins, R.F. & Borm, P.J.A. (1994) Effects of aramid, a high strength synthetic fiber, on respiratory cells *in vitro*. *Drug Chem. Toxicol.*, 17, 75–92

McGavran, P.D., Moore, L.B. & Brody, A.R. (1990) Inhalation of chrysotile asbestos induces rapid cellular proliferation in small pulmonary vessels of mice and rats. *Am. J. Pathol.*, 136, 695–705

Metcalf, R.A., Welsh, J.A., Bennett, W.P., Seddon, M.B., Lehman, T.A., Pelin, K., Linnainmaa, K., Tammilehto, L., Mattson, K., Gerwin, B.I. & Harris, C.C. (1992) p53 and Kirsten-ras mutations in human mesothelioma cell lines. *Cancer Res.*, 52, 2610–2615

Moalli, P.A., MacDonald, J.L., Goodlick, L.A. & Kane, A.B. (1987) Acute injury and regeneration of the mesothelium in response to asbestos fibers. *Am. J. Pathol.*, 128, 426–445

Mossman, B.T., Craighead, J.E. & MacPherson, B.V. (1980) Asbestos-induced epithelial changes in organ cultures of hamster trachea: inhibition by retinyl methyl ether. *Science*, 207, 311–313

Mossman, B.T., Marsh, J.P. & Shatos, M.A. (1986) Alteration of superoxide dismutase activity in tracheal epithelial cells by asbestos and inhibition of cytotoxicity by antioxidants. *Lab Invest.*, 54, 204–212

Mossman, B.T., Janssen, Y.M.W., Marsh, J.P., Sesko, A., Shatos, M.A., Doherty, J., Adler, K.B., Hemenway, D., Mickey, R., Vacek, P., Petruska, J. & Kagan, E. (1991) Development and characterization of a rapid onset rodent inhalation model of asbestosis for disease prevention. *Toxicol. Pathol.*, 19, 412–418

Noble, P.W., Henson, P.M. & Riches, D.W.H. (1991) Insulin-like growth factor-1 mRNA expression in bone marrow derived macrophages is stimulated by chrysotile asbestos and bleomycin. *Chest*, 99, S79–S80

Nuorva, K., Makitaro, R., Huhta, E., Kamel, D., Vahakangas, K., Bloigi, R., Soini, Y. & Paakko, P. (1994) p53 Protein accumulation in lung carcinomas of patients exposed to asbestos and tobacco smoke. *Am. J. Respir. Crit. Care Med.*, 150, 528–533

Oberdörster, G. (1994). Extrapolation of results from animal inhalation studies with particles to humans. In: Dungworth, D., Mohr, U., Mauderly, J. & Oberdorster, G., eds, *Toxic and Carcinogenic Effects of Solid Particles in the Respiratory Tract*, Washington, DC, ILSI Press, pp. 57–73

Oghiso, Y. & Kubota, Y. (1986) Interleukin 1-like thymocyte and fibroblast activating factors from rat alveolar macrophages exposed to silica and asbestos particles. *Jpn. J. Vet. Sci.*, 48, 461–471

Pegg, A.E. (1988) Polyamine metabolism and its importance in neoplastic growth and as a target for chemotherapy. *Cancer Res.*, 48, 759–744

Perkins, R.C., Sheule, R.K. & Holian, A. (1991) *In vitro* bioactivity of asbestos for the human alveolar macrophage and its modification by IgG. *Am. J. Respir. Cell. Mol. Biol.*, 4, 532–537

Perderiset, M., Marsh, J.P. & Mossman, B.T. (1991) Activation of protein kinase C by crocidolite asbestos in hamster tracheal epithelial cells. *Carcinogenesis*, 12, 1499–1502

Quinlan, T.R., BéruBé, K.A., Marsh, J.P., Janssen, M.W., Taishi, P., Leslie, K.O., Hemenway, D., O'Shaughnessy, P.T., Vacek, P. & Mossman, B.T. (1995) Patterns of inflammation, cell proliferation, and related gene expression in lung after inhalation of chrysotile asbestos. *Am. J. Pathol.*, 147, 728–738

Rajan, K.T., Wagner, J.C. & Evans, P.H. (1972) The response of human pleura in organ culture to asbestos. *Nature*, 238, 346–347

Resnick, D., Freedman, N.J., Xu, S. & Krieger, M. (1993) Secreted extracellular domains of macrophage scavenger receptors form elongated trimers which specifically bind crocidolite asbestos. *J. Biol. Chem.*, 268, 3538–3545

Rom, W.N. (1991) Relationship of inflammatory cell cytokines to disease severity in individuals with occupational inorganic dust exposure. *Am. J. Ind. Med.*, 19, 15–27

Rom, W.N., Travis, W.D. & Brody, A.R. (1991) Cellular and molecular basis for the absestos-related diseases. *Am. Rev. Respir. Dis.*, 143, 408–422

Roney, P.L. & Holian, A. (1989) Possible mechanism of chrysotile asbestos-stimulated superoxide production in guinea pig alveolar macrophages. *Toxicol. Appl. Pharmacol.*, 100, 132–144

Rosenthal, G.J., Germolec, D.R., Blazka, M.E., Corsini, E., Simeonova, P., Pollock, P., Kong, L.-Y., Kwon, J. & Luster, M.I. (1994) Asbestos stimulates IL-8 production from human lung epithelial cells. *J. Immunol.*, 153, 3237–3244

Rutten, A.A.J.J.L., Bermudez, E., Mangum, J.B., Wong, B.A., Moss, O.R. & Everitt, J.I. (1994) Mesothelial cell proliferation induced by intrapleural instillation of man-made fibers in rats and hamsters. *Fundam. Appl. Toxicol.*, 23, 107–116

Rutten, A.A.J.J.L., Bermudez, E., Stewart, W., Everitt, J. & Walker, C. (1995) Expression of insulin-like growth factor II in spontaneously immortalized rat mesothelial and spontaneous mesothelial cells: a potential autocrine role of insulin-like growth factor II. *Cancer Res.*, 55, 3534–3639

Scheule, R.K. & Holian, A. (1989) IgG specifically enhances chrysotile asbestos-stimulated superoxide anion production by the alveolar macrophage. *Am. J. Respir. Cell. Mol. Biol.*, 1, 313–318

Scheule, R.K. & Holian, A. (1990) Modification of asbestos bioactivity for the alveolar macrophage by selective protein adsorption. *Am. J. Cell. Mol. Biol.*, 2, 441–448

Schreck, R., Rieber, P. & Baeuerle, P.A. (1991) Reactive oxygen intermediates as apparently widely used messengers in activation of NFκB transcription facto and HIV. *EMBO J.*, 10, 2247–2258

Sesko, A.M. & Mossman, B.T. (1989) Sensitivity of hamster tracheal epithelial cells to asbestiform minerals is modulated by serum and by transforming growth factor. *Cancer Res.*, 49, 2743–2749

Sesko, A., Cabot, M. & Mossman, B. (1990) Hydrolysis of inositol phospholipids precedes cellular proliferation in asbestos-stimulated tracheobronchial epithelial cells. *Proc. Natl Acad. Sci. USA*, 87, 7385–7389

Shatos, M.A., Doherty, M.A., Marsh, J.P. & Mossman, B.T. (1987) Prevention of asbestos-induced cell death in rat lung fibroblasts and alveolar macrophages by scavengers of active oxygen species. *Environ. Res.*, 44, 103–116

Shull, S., Heintz, N.H., Periasamy, M., Manohar, M., Janssen, Y.M.W., Marsh, J.P. & Mossman, B.T. (1991) Differential regulation of antioxidant enzymes in response to oxidants. *J. Biol. Chem.*, 266, 24398–24340

Simeonova, P.P. & Luster, M.I. (1995) Iron and reactive oxygen species in the asbestos-induced tumor necrosis factor α response from alveolar macrophages. *Am. J. Respir. Cell. Mol. Biol.*, 12, 676–683

Timblin, C.R., Janssen, Y.W.M. & Mossman, B.T. (1995) Transcriptional activation of the proto-oncogene c-jun by asbestos and H_2O_2 is directly related to increased proliferation and transformation of tracheal epithelial cells. *Cancer Res.*, 55, 2723–2726

Van der Meeren, A., Seddon, M.B., Betsholtz, C.A., Lechner, J.F. & Gerwin, B.I. (1993) Tumorigenic conversion of human mesothelial cells as a consequence of platelet-derived growth factor-A chain overexpression. *Am. J. Respir. Cell. Mol. Biol.*, 8, 214–221

Versnel, M.A., Hagemeijer, A., Bouts, M.J., van der Kwast, T.H. & Hoogsteden, H.C. (1988) Expression of PDGF B-chain and A-chain genes in ten malignant mesothelioma cell lines derived from primary and metastatic tumors. *Oncogene*, 2, 601–605

Versnel, M.A., Claessen-Welsh, L., Kammacher, A., Bouts, M.J., van der Kwast, T.H., Eriksson, A., Willemsen, R., Weima, S.N., Hoogsteden, H.C., Hagemiejer, A. & Heldin, C.H. (1991) Human malignant mesothelioma cell lines express PDGF-β-receptors whereas cultured normal mesothelial cells express predominantly PDGF-α-receptors. *Oncogene*, 6, 2005–2011

Walker, C., Bermudez, E., Stewart, W., Bonner, J., Molloy, C.J. & Everitt, J. (1992) Characterization of platelet-derived growth factor and platelet-derived growth factor receptor expressn in asbestos-induced rat mesothelioma. *Cancer Res.*, 52, 301–306

Walker, C., Everitt, J., Ferriola, P.A., Stewart, W., Mangum, J. & Bermudez, E. (1995) Autocrine growth factor stimulation by transforming growth factor α in asbestos-induced transformed rat mesothelial cells. *Cancer Res.*, 55, 530–536

Warheit, D.B., Chang, L.Y., Hill, L.H., Snyderman, R. & Brody, A.R. (1985) Inhaled asbestos activates complement-dependent chemoattractants for macrophages. *Lab. Invest.*, 52, 505–514

Weitzman, S.A. & Graceffa, P. (1984) Asbestos catalyzes hydroxyl and superoxide radical generation from hydrogen peroxide. *Arch. Biochem. Biophys.*, 228, 373–376

Weitzman, S.A. & Weitberg, A.B. (1985) Asbestos-catalysed lipid peroxidation and its inhibition by desferroxamine. *Biochem J.*, 225, 259–262

Zanella, C.L., Tritton, T.R., Posada, J. & Mossman, B.T. (1995) Involvement of mitogen-activated protein kinases (MAPK) in proliferative responses to asbestos. *Proc. Am. Assoc. Cancer Res.*, 36, 54

Zhang, Y., Lee, T.C., Guillemin, B., Yu, M.-C. & Rom, W.N. (1993) Enhanced IL-1β and tumor necrosis factor α release and messenger RNA expression in macrophages from idiopathic pulmonary fibrosis or after asbestos exposure. *J. Immunol.*, 150, 4188–4196

Corresponding author
K.E. Driscoll
The Procter & Gamble Company, Human Safety Department, P.O. Box 398707, Cincinnati, OH 45253, USA

Short-term animal studies for detecting inflammation, fibrosis and pre-neoplastic changes induced by fibres

K. Donaldson

The present review is aimed at describing biological endpoints in short-term animal studies; the majority of these studies span a period of days to weeks following exposure, and the endpoints are those that precede end-stage pathological change. The review will begin by describing the range of endpoints that have been used in short-term animal studies of the biological effects of fibres. The main focus of the review will be the detailed role of free radicals, cytokines and growth factors in pre-neoplastic and pre-fibroplastic change.

Species and exposure routes
The animal species most commonly used are rats, mice and hamsters and the routes of exposure are inhalation, intratracheal instillation and intraperitoneal instillation.

Endpoints in short-term animal studies
Short-term animal studies have been used for a wide number of purposes and some of these endpoints are listed in Tables 1–3.

Short-term animal studies on the airways and lung parenchyma
Inhalation exposure has been used frequently and to great effect in the elegant and influential series of studies carried out by A.R. Brody and co-workers (e.g. Brody et al., 1981). These experiments demonstrated the site of deposition and the early macrophage and tissue responses to chrysotile asbestos. Instillation of fibres has also been used as a less expensive and more rapid alternative to inhalation. However, the likely differences in the site of deposition and relatively massive dose delivered in a single liquid bolus by instillation, means that responses may differ from those caused by a gradual build-up of inhaled dose. The rat has been used most commonly as the study animal in these experiments.

Inflammation and cytokines
Numerous studies have demonstrated the ability of asbestos exposure to cause inflammation as assessed by bronchoalveolar lavage (BAL) techniques (reviewed in Donaldson & Brown, 1993). However, there have been fewer studies with other respirable fibres. Warheit (1993) has shown modest non-persistent inflammation with wollastonite and *para*-aramid fibres after five days of exposure at 613–1344 fibres/mL *para*-aramid and 123–835 fibres/mL wollastonite; in contrast, crocidolite-exposed rats did show persistent inflammation, as judged by neutrophil influx. Also, studies of inflammation caused by three respirable fibres – long amosite, glass fibre (GF) 100/475 and silicon carbide fibres at 1000 fibres/mL – were carried out for up to 14 days by the author in the Colt Fibre Programme of the Institute of Occupational Medicine (K. Donaldson, unpublished data). All of the fibres caused some degree of inflammation, but there was no clear discrimination between GF 100/475, which was non-pathogenic, and the other fibres, which did cause pathology. Unfortunately, no studies of the persistence of the inflammation were carried out, which might have provided better discrimination (see Warheit et al., 1994).

The increased release of cytokines and factors mitogenic for fibroblasts by BAL macrophages from asbestos-exposed rats has been described in many short-term studies (reviewed in Donaldson & Brown, 1993). Few studies have involved characterization of the cytokines. Li et al. (1993a) described sustained increases in tumour necrosis factor α (TNFα) and interleukin (IL)-1 release from BAL cells after instillation of crocidolite. Recently, Brody (1995)

Table 1. Short-term studies on lung parenchyma and airways		
Endpoint	**References**	**Mode[a]**
Pattern and site of deposition	Brody et al., 1981; Pinkerton et al., 1986	INH
Mechanism of macrophage recruitment to site of fibre deposition	Warheit et al., 1984	INH
Permeability of the bronchiolo-alveolar epithelium	Folkesson et al., 1993	IT
Up-regulation of transcription of proto-oncogenes and other transcription-related genes	Quinlan et al., 1994	INH
Induction of antioxidant defence	Janssen et al., 1991; Quinlan et al., 1994	INH
Proliferation in terminal bronchiolar region		
• asbestos	Brody & Overby, 1989; Warheit, 1994	INH
• comparing asbestos and man-made fibres	K. Donaldson, unpublished	INH
Morphometry of early lesions	Chang et al., 1988	INH
Generation of free radicals in the lung	Schapira et al., 1994	IT
Inhibition of lung injury by antioxidant	Mossman et al., 1990	INH
Fibre retention: exacerbating role of oxidants	Pinkerton et al., 1989	INH
	McFadden et al., 1986	IT
BAL leukocyte kinetics	Donaldson et al., 1988a	INH
Loss of phagocytic capacity of the BAL cells	Donaldson et al., 1990	INH
BAL as a short-term bioassay for predicting fibre toxicity	Warheit, 1994	INH

[a]Mode, mode of delivery of fibre to the lung: INH, inhalation; IT, intratracheal instillation.
BAL, bronchoalveolar lavage.

exposed rats to chrysotile asbestos and demonstrated increased immunochemical staining of transforming growth factor α (TGFα) in clara cells, type II epithelial cells and macrophages when compared to lungs following inhalation of iron beads. There are no known studies involving the comparison of asbestos with other fibres or the role of fibre length in cytokine release following short-term inhalation, but these are clearly warranted.

Free radicals
The evidence of a role for free radicals in the pathogenicity of fibres comes mostly from the study of asbestos. In these studies, the whole lung is exposed and markers of oxidative stress are detected, and it is therefore difficult to untangle whether the source of the oxidative stress is the fibres themselves or the inflammatory leukocytes that are recruited to the site of deposition. Mossman and coworkers (Quinlan et al., 1994) examined the expression of mRNA for manganese superoxide dismutase (MnSOD) and copper-zinc superoxide dismutase (Cu-ZnSOD) in lungs of rats exposed to 60 or 2800 fibres/mL crocidolite. Levels of mRNA for MnSOD and Cu-ZnSOD were increased with both exposure levels; induction was faster and greater in magnitude with the high dose. The contribution of inflammatory cells was not specifically addressed in this study but SOD mRNA was increased at the low exposure level when there was no neutrophil influx in the BAL. This suggests that neutrophils are not directly involved in the oxidative stress. Production of oxidants by activated macrophages remains a possibility, but the data showing the production of free radicals by BAL cells in animals exposed to fibres are complex and conflicting (reviewed by Kamp et al.,1992) and need to be addressed systematically.

Mossman et al. (1990) produced compelling evidence that oxidative stress was important in asbestos fibrosis by intervening with the antioxidant enzyme catalase and ameliorating the fibrogenicity and inflammogenicity of inhaled asbestos. The fibre used was NIEHS crocidolite (average length, 11.5 μm) at 10 mg/m^3 for up to 20 days. In these studies, rats were fitted subcutaneously with minipumps delivering a 'long-life' version of the antioxidant enzymes catalase and SOD; SOD had no effect. Whilst the highest dose of catalase (8000

Table 2. Short-term studies on the pleural cavity response to fibres

Endpoint	Reference	Mode[a]
Dose to the pleura	Gelzleichter et al., 1996	INH
Mesothelial cell proliferation	Adamson et al., 1993	IT
	Coin et al., 1991	INH
Pleural leukocyte population changes	Oberdörster et al., 1983; Li et al., 1991	IT
Pleural leukocyte cytokines	Li et al., 1992	IT
Pleural leukocyte release of anti-fibrinolytic activity	Li et al., 1991	IT

[a]Mode, mode of delivery of fibre to the lung: INH, inhalation; IT, intratracheal instillation.

international units/day) inhibited the increase in hydroxyproline seen in the asbestos-exposed group, it also lowered the total number of macrophages and neutrophilic cells in lavage; this level of catalase had no effect on protein or alkaline phosphatase compared to the asbestos control. These data strongly support a role for hydrogen peroxide in mediating and enhancing the harmful effects of asbestos, and increased mRNA for glutathione peroxidase has also been detected in asbestos fibre-exposed rat lung (Janssen et al., 1991), which suggests further a need for the fibre-exposed lung to deal with increased peroxides.

Hydroxyl radicals (OH$^\bullet$) are one of the most reactive of the free radicals that arise in biological systems, and there is evidence for the formation of OH$^\bullet$ in lungs of rats following fibre instillation (Schapira et al., 1994). In this study, the reaction between OH$^\bullet$ and salicylate was used to trap OH$^\bullet$ and the product of the reaction was quantified; fourfold to 10-fold more of this product was found in the lungs of rats instilled with 'iron-loaded' chrysotile with no details of the size distribution.

The problem remains that, in the case of exposure of a whole lung, it is difficult to dissect out the source of the oxidant – is it inflammatory cell- or fibre surface-derived? However, in the case of carcinogenic change, the recent studies of K.E. Driscoll may provide a strategy that could tease out the role of inflammatory cells. Driscoll et al., (1996) revealed that exposure of rats to silica can be associated with a detectable rate of mutation in the hprt locus of the alveolar type II epithelial cells; furthermore, the inflammatory cells themselves are probably responsible for the mutation (Driscoll et al., 1996). Because asbestos or other fibres are generally considered to be more carcinogenic than silica, mutations might be detectable using this technique with short-term high-fibre exposures; such studies could be an important approach to dissecting the source of mutagenic reactive oxygen intermediates in the carcinogenic response to fibres.

Table 3. Short-term studies on fibres in the peritoneal cavity

Endpoint	Reference
Induction of fibrinolytic activity in macrophages	Hamilton, 1980
Iron metabolism in macrophages	Koerten et al., 1986
Mesothelial cell injury/proliferation	Moalli et al., 1987
Induction of angiogenesis	Branchaud et al., 1989
Macrophage activation	Donaldson et al., 1982
Inflammatory response	Donaldson et al., 1988b; Branchaud et al., 1993
Effects of fibre dissolution	Donaldson et al., 1994
Distribution of fibres in the peritoneal cavity	Morgan et al., 1993

Table 4. BrdU incorporation and pathology in rats exposed to different fibres

Fibre	Exposure	Proliferation	Pathology[a]
GF 100/475	1000 f/mL, 7 h	–/+	11/0/0
Long fibre amosite	1000 f/mL, 7 h	++++	38/17/5
Silicon carbide	1000 f/mL, 7 h	++++	24/12/24

[a]Percentage of all tumours/percentage of all cancers/percentage of mesotheliomas.
GF, glass fibre.

In addition to the effects described above, oxidants could contribute to the pathogenic effects of fibres in indirect ways, by enhancing the dose via increased retention of fibres. A. Churg and co-workers (McFadden et al., 1986) demonstrated increased penetration of amosite asbestos into the airway walls of guinea-pigs exposed to cigarette smoke by inhalation compared to guinea-pigs exposed to asbestos alone by instillation. This phenomenon may be related to that of the increased asbestos retention detected in the lungs of rats exposed to asbestos and ozone compared to rats exposed to asbestos alone (Pinkerton et al., 1989). Further research is necessary to dissect out the factors responsible for this oxidant stress-mediated inhibition of clearance.

Proliferation

Because pulmonary fibrosis and cancer are the result of overgrowths of tissue, attention has been given to the detection of proliferation in fibre-exposed lungs. Brody and co-workers, in pioneering studies, described the highly focal nature of deposition and the resulting proliferative and tissue response arising in the short-term from the deposition of chrysotile fibres. The bronchiolar/alveolar deposition of fibres, accumulation of macrophages and the developing increase in tissue mass, quantified at the first alveolar duct bifurcation (Chang et al., 1988; Warheit et al., 1994), were shown to be a result of increased proliferation of the epithelial and interstitial cells in this area (e.g. Brody & Overby, 1989). Furthermore growth factors such as platelet-derived growth factor (PDGF), insulin growth factor-1 (IGF) and TGFβ, released by particle-exposed macrophages, were hypothesized to be the principal stimuli for this proliferation (Brody, 1991).

Adamson and Bowden (1987a,b) utilized instillation of short and long crocidolite asbestos in mice to study the relationship of proliferation to fibrosis. The short fibres caused a transient inflammation and proliferation in various compartments and there was no long-standing pathological change. In contrast, the long fibres caused sustained inflammation, epithelial injury leading to intersititialization of fibres and an accompanying continued stimulus to epithelial proliferation, as well as cell division in the interstitium, culminating in fibrosis.

In addition to the likely role of macrophage-derived cytokines, fibres themselves may stimulate proliferative genes directly (Janssen et al., 1995). B.T. Mossman and co-workers (Quinlan et al., 1994) demonstrated that mRNA for cell proliferation-associated genes such as ornithine decarboxylase and c-*jun* is increased in rats exposed to 2800 fibres/mL crocidolite asbestos. Recent in-vitro data suggest that oxidative stress caused by fibres (as manifested by a disturbance to the glutathione: glutathione disulfide (GSH:GSSG) redox balance in the cell) is a key feature in leading to the activation of transcription of the activator protein-1 (AP-1) complex (Janssen et al., 1995), which provides a link between free radicals and proliferation. In linking free radical stress and proliferation, this brings together the two most compelling hypotheses on the mechanisms of fibre pathology into a testable hypothesis that could be addressed in short-term inhalation studies, such as the depletion of intracellular GSH by treating rats with buthionine sulfoximine.

Furthermore, there is every reason to suspect that cytokines and the direct effects of fibres may act synergistically on the GSH:GSSG axis. We have demonstrated that TNFα depletes intracellular GSH in alveolar epithelial cells (Rahman et al., 1995); therefore, any epithelial cell both in contact with a fibre and also exposed to TNFα from a nearby macrophage that has phagocytozed a fibre could be the target of a double GSH-depleting effect.

Certain cytokines, such as TNFα and TGFβ, are able to cause sustained expression of c-*fos* and c-*jun* and may be possible tumour promoters (Angel & Karin, 1991). Continued stimulation of expression of the components of the AP-1 complex and nuclear factor κB (NFκB), or activation of MAPKinase by cytokines, together with fibres, could cause a sustained expression of the cytokines leading to conditions that could be very favourable for proliferation, fibrosis and cancer.

Proliferation in short-term animal studies with fibres other than asbestos
Few studies have addressed the ability of fibres other than asbestos to cause proliferation. However, data from the Colt Fibre Programme of the Institute of Occupational Medicine are now available on the three fibre types as described in Table 4. The results, outlined in the table, revealed that a 7-h exposure to 1000 fibres/mL was associated with increased 5-bromo-2´deoxyuridine (BrdU) incorporation in two fibres that caused long-term pathology, but no increased proliferation was present in the rats inhaling GF 100/475, which was not pathogenic.

Warheit (1993) examined the proliferative response to inhaled wollastonite and *para*-aramid fibres for five days at 613-1344 fibres/mL *para*-aramid and 123-835 fibres/mL wollastonite. There were no significant increases in the proportions of proliferating cells over sham-exposed rats with either of these fibres one week or one month after the end of exposure. Wollastonite is a relatively non-pathogenic fibre in long-term inhalation studies (McConnell *et al.*, 1991). *para*-Aramid inhalation was found to cause carcinomas in a two-year study (Lee *et al.*, 1988); however, these tumours were later re-evaluated and the majority of pathologists no longer describe these lesions as carcinomas. In a further series of experiments, Warheit *et al.* (1994) compared chrysotile asbestos and *para*-aramid at equal fibre numbers. Only a very transient burst of proliferation was found with *para*-aramid, while the chrysotile-exposed rats showed a sustained proliferative response at three months after exposure. It would appear, therefore, that pathogenic fibres may have the ability to cause persistent proliferation in short-term inhalation studies; this may form the basis of a short-term predictive test (Warheit, 1993). More fibres need to be studied to pursue this promising area of short-term testing.

Short-term animal studies on the pleura
Because of the association between asbestos and mesothelioma (and other forms of pleural pathology), attention has been focused on the pleural mesothelium as a target following exposure to asbestos in short-term animal studies. The target tissues are the mesothelium and the pleural leukocytes, with the likely events being: (i) the interaction between fibre and mesothelial cells leading to fibrosis and mesothelioma; and (ii) the interaction of fibres with pleural leukocytes, causing them to release cytokines and oxidants that could have proliferative or mutagenic effects on the mesothelial cells.

Inflammatory responses in the pleural space
K. Donaldson, X. Li and co-workers made use of a rat model instilling 1 mg (up to 5 mg) of UICC crocidolite asbestos into the lungs and assessing various features of the pleural leukocyte response. The results from these studies are summarized in Table 5.

It was notable that all of the above effects were found in the absence of any fibres in the pleural

Table 5. Effect of intratracheal instillation of crocidolite on the pleural leukocyte response

Pleural leukocyte response	Reference
Change in the population	Li *et al.*, 1991
Increased production of plasminogen activator inhibitor	Li *et al.*, 1991
Lowered production of TNFα	Li *et al.*, 1992
Initial lowering then stimulation of IL-1 activity	Li *et al.*, 1993b
Detachment injury to mesothelial cells *in vitro* by pleural leukocytes	Li *et al.*, 1994

TNFα, tumour necrosis factor α; IL-1, interleukin-1.

Table 6. An assessment of the information available from short-term studies in the following compartments[a]

	BAL/lung		Pleura		Peritoneal cavity	
	Asbestos	MMVF	Asbestos	MMVF	Asbestos	MMVF
Inflammation/cytokine	2/3	1	2	1	2	1
Free radicals	2	0	0	0	2	0
Proliferation	2/3	1	1	1	1	0
Proto-oncogenes/tumour suppressor genes	1	0	0	0	0	0

[a]Data are evaluated on a scale of 0 to 5, where 0 = no information available and 5 = substantial information available.
BAL, bronchoalveolar lavage; MMVF, man-made vitreous fibres.

leukocytes, despite intensive efforts to find fibres or a route for fibres from the bronchoalveolar space to the pleura (Li et al., 1993c). We conclude that changes in the pleural leukocytes can take place without interaction of fibres with the pleural leukocytes. Such effects may arise from interaction of fibres with the mesothelial cells, presumably from the basal side in the first instance, as suggested by Adamson et al. (1993). Oberdörster (1983) described sustained changes in the pleural leukocyte population following instillation of amosite into the bronchoalveolar space and again no evidence of fibres in the leukocytes. These population changes were characterized by an influx of monocytic cells, suggesting that there is generation of a chemotactic gradient in the pleural space; the origin of this gradient is not known. Further studies are required to examine the production of chemotaxins by pleural leukocytes and mesothelial cells after inhalation exposure; it is not clear whether there is substantial translocation of fibres to the pleural leukocytes to stimulate chemotaxin release, but the studies of Li et al. (1991, 1992, 1993b, 1994) and others (Oberdörster, 1983; Adamson et al., 1993) strongly suggest that, even when a large number of fibres are deposited in the lung, there is little, if any, sign of fibre translocation to the pleural leukocytes.

Proliferation in the pleura
Because of the importance of the mesothelium as a target for the pathogenic effects of fibres, the proliferative responses in the pleura following fibre exposure may be an important index of potential to produce pathology in the long-term. Brody and coworkers (Coin et al., 1991) found modest increases in proliferation in the pleural mesothelium following a single high-dose inhalation exposure to chrysotile in mice. In the studies of Warheit et al. (1994) that compared *para*-aramid and chrysotile after a two-week exposure at equal fibre number, there was a clear stimulation of proliferation in the pleural tissue with chrysotile but none with *para*-aramid.

In the elegant studies of Adamson et al. (1993), long and short crocidolite were instilled into mouse lungs and the proliferative responses measured. There was marked interstitial, bronchiolar and bronchial proliferation with long fibres compared to very modest proliferation with the short fibres. This same difference was also noted in the mesothelial proliferative response, demonstrating the potency of the long fibres in causing mesothelial proliferation. It was notable that Adamson et al. (1993) concluded that there was no transfer of fibres to the pleura and that diffusable mitogenic 'signals' from the underlying fibre-containing parenchymal macrophages must be responsible for the wave of proliferation (see above).

Adamson et al. (1994) went on to demonstrate, however, that a number of instilled materials – silica, bleomycin, as well as hyperoxia – were capable of stimulating a wave of mesothelial proliferation. This proliferation was sustained with silica and bleomycin to the same degree as that found in mice instilled with long crocidolite. Since neither silica nor bleomycin are associated with mesothelioma or the degree of pleural pathology shown by asbestos, the exact meaning of this proliferative response is difficult to establish. However, a true comparison in rats following inhalation and instillation of a range of materials including fibres is necessary to compare the potency of these materials and to determine whether this potentially promising strategy for predictive testing has application.

Short-term studies in the peritoneal cavity

The peritoneal cavity has been an area of study for several reasons: (i) there is a high proportion of peritoneal mesotheliomas in some human population studies; (ii) the rat peritoneal mesothelioma test is one that has received substantial interest as a screening test for carcinogenic fibres; (iii) the peritoneal cavity is a useful site for modelling pleural response, since it shows all of the responses found in the pleura – inflammation, fibrosis and mesothelioma; and (iv) it is easy to deliver a defined dose of fibres to the peritoneal mesothelium.

The drawbacks of this mesothelial site are well known and relate to the absence of a clearance system and filtration system, which undoubtedly dramatically attenuates the number and size range of fibres that penetrate to the pleural mesothelium. Additionally, inhaled fibres are not deposited on the pleural mesothelial surface, unlike the situation that arises following instillation in the peritoneal cavity.

Inflammation in the peritoneal cavity in response to fibres

Hamilton (1980) initially utilized the mouse peritoneal cavity to demonstrate the potential of fibres to cause inflammation after exposure of serosal surfaces, as shown by induction of plasminogen activator stimulation in the peritoneal exudate cells. Donaldson et al. (1982) went on to demonstrate activation of mouse peritoneal macrophages by instilled asbestos, as shown by increased Fc receptor expression but not by full activation to tumoricidal status. The activation of macrophages in this site by asbestos fibres was associated with systemic immunosuppression (Hannant et al., 1985; Szymaniec et al., 1990), which could be related to the release of a lympho-suppressive factor by the asbestos-activated peritoneal macrophages (Donaldson et al., 1984). Kane and co-workers went on to dissect the detailed cellular events underlying inflammation in the mouse peritoneal cavity, emphasizing the role of free radicals in macrophage and mesothelial injury and regeneration (see below). The inflammatory response in the peritoneal cavity is acutely responsive to fibres and is relatively unresponsive to particles, with titanium dioxide (TiO_2) and even silica having very little effect (Donaldson et al., 1988; Branchaud et al., 1993). There is also very little response to short amosite while long amosite causes severe inflammation (Donaldson et al., 1989).

Free radicals

Donaldson and Cullen (1984) described increased release of superoxide anion by macrophages from the peritoneal cavity of mice instilled with UICC chrysotile asbestos, suggesting that release of ROS could play a role in the long-term pathology found in the peritoneal cavity with asbestos instillation. This was confirmed by Goodglick and Kane (1990), who noted amelioration of the cell injury (as assessed by Trypan blue staining of

Table 7. Rational application of different exposure routes in short-term studies of fibres		
Lung	**Pleura**	**Peritoneal cavity**
Endpoints measured after **inhalation** exposure are potentially useful for risk assessment.	Endpoints measured after **inhalation** exposure are potentially useful for risk assessment.	Potentially suitable for mechanism studies on mesothelioma, pleural inflammation/ fibrosis.
Endpoints measured after **instillation** exposure are potentially useful for mechanism studies on fibrosis/lung cancer.	Endpoints measured after **instillation** exposure are potentially useful for mechanism studies on mesothelioma/ pleural fibrosis.	

Table 8. Criticisms of the various types of short-term study with fibres		
Inhalation	**Instillation into the lung**	**Peritoneal cavity instillation**
Relatively high exposures used, except in a few cases.	Instillation causes sudden inflammation unlike the effect of gradual build-up found with inhalation.	Instillation causes sudden inflammation unlike the effect of gradual build-up found with inhalation.
	Presence of fluid vehicle may synergize with the effect of fibres.	Bolus injection associated with systemic immunosuppression.
	No filtration of very long or nonrespirable fibre types.	Nonphysiological deposition directly onto the mesothelium.

diaphragm and levels of lactic dehydrogenase (LDH) in the peritoneal lavage fluid) caused in the peritoneal cavity by asbestos after co-injection of SOD, catalase and the iron chelator desferrioxamine (Goodglick & Kane, 1990).

Proliferation
Kane and co-workers (Moalli et al., 1987) described the proliferative response of the mesothelium following instillation of long and short crocidolite asbestos. There was more than a 10-fold increase in labelled cells with long amosite treatment and the wave of proliferation was sustained for longer than 14 days; the response tailed off rapidly in the case of short amosite. The authors concluded that both direct fibre injury and inflammatory mediators could be responsible for mesothelial injury, and that macrophage growth factors were responsible for the mesothelial proliferation.

Fibres other than asbestos
When a large range of fibres, including refractory ceramic fibres (RCF) and man-made vitreous fibres (MMVF), were tested at equal fibre numbers for their ability to cause inflammation in the peritoneal cavity, there was found to be very little difference between them (Donaldson et al., 1994). This may be because of the acute sensitivity of the mesothelial cells to damage by long fibres or the failure of long fibres to clear from the peritoneal cavity (Moalli et al., 1987). However, since residence in the lung can lead to changes in the structure of more soluble fibres, it was hypothesized that MMVF (soluble) might lose some of their biological activity when incubated in buffer, while RCF and amphibole fibres (insoluble) might retain their biological activity. When the fibres were treated to mimic residence in the lung by placing them in buffer, there was a loss of inflammogenic potential in the case of the MMVF, but none in the case of the RCF (Donaldson et al., 1994) or amphibole asbestos fibres (Donaldson, 1994). It was suggested that treatment to mimic residence in the lung prior to use in short-term assays may yield a better index of the potential of fibres to cause long-term pathology (Donaldson, 1994).

Conclusion
The conclusions and limitations of short-term animal studies are presented in Tables 6–8.

References
Adamson, I.Y.R. & Bowden, D.H. (1987a) Response of mouse lung to crocidolite asbestos. I. Minimal fibrotic reaction to short fibres. *J. Pathol.*, 152, 99–107

Adamson, I.Y.R. & Bowden, D.H. (1987b) Response of mouse lung to crocidolite asbestos. II. Pulmonary fibrosis after long fibres. *J. Pathol.*, 152, 109–117

Adamson, I.Y.R., Bakowska, J. & Bowden, D.H. (1993) Mesothelial cell proliferation after installation of long or short asbestos fibers into the mouse lung. *Am. J. Pathol.*, 43, 149–153

Adamson, I.Y.R., Bakowska, J. & Bowden, D.H. (1994) Mesothelial cell proliferation: a non-specific response to lung injury associated with fibrosis. *Am. Rev. Respir. Cell Mol. Biol.*, 10, 253–258

Angel, P. & Karin, M. (1991) The role of jun, fos and the AP-1 complex in cell-proliferation and transformation. *Biochim. Biophys. Acta*, 1072, 129–157

Branchaud, R.M., MacDonald, J.L. & Kane, A.B. (1989) Induction of angiogenesis by intraperitoneal injection of asbestos. *FASEB J.*, 3, 1747–1752

Branchaud, R.M., Garant, L.J. & Kane, A.B. (1993) Pathogenesis of mesothelial reactions to asbestos fibres – recruitment and macropahge activation. *Pathobiology*, 61, 154–163

Brody, A.R. (1991) Production of cytokines by particle-exposed macrophages. In: *Cellular and Molecular Aspects of Fiber Carcinogenesis*, Cold Spring Harbour, CSH Press, pp. 83–102

Brody, A.R. (1995) The role of growth factors in fibroproliferative disease: in vitro models and in vivo correlates. Paper presented at the 5th International Inhalation Symposium, Hannover. February 1995.

Brody, A.R. & Overby, L.H. (1989) Incorporation of tritiated thymidine by epithelial cells in broncho-alveolar regions of asbestos-exposed rats. *J. Pathol.*, 134, 133–140

Brody, A.R., Hill, L.H., Adkins, B. & O'Connor, R.W. (1981) Chrysotile asbestos inhalation in rats: deposition pattern and reaction of alveolar epithelium and pulmonary macrophages. *Am. Rev. Respir. Dis.*, 123, 670–679

Chang, L.-Y., Overby, L.H., Brody, A.R. & Crapo, J.D. (1988) Progressive lung cell reactions and extracellular matrix production after a brief exposure to asbestos. *Am. J. Pathol.*, 131, 156–169

Coin, P.G., Moore, L.B., Roggli, V. & Brody, A.R. (1991) Pleural incorporation of ^{3}H-TdR after inhalation of chrysotile asbestos in the mouse. *Am. Rev. Respir. Dis.*, 143, A604

Donaldson, K. (1994) Biological activity of respirable indistrial fibres treated to mimic residence in the lung. *Toxicol. Lett.*, 72, 229–305

Donaldson, K. & Brown, G.M. (1993) Bronchoalveolar lavage in the asessment of the cellular response to fiber exposure. In: Warheit, D.B., ed., *Fiber Toxicology*, New York, Academic Press, pp. 117–138

Donaldson, K. & Cullen, R.T. (1984) Chemiluminescence of asbestos-activated macrophages. *Br. J. Exp. Pathol.*, 65, 81–90

Donaldson, K. Davis, J.M.G. & James, K. (1982) Characteristics of peritoneal macrophages induced by asbestos injection. *Environ. Res.*, 29, 414–424

Donaldson, K., Davis, J.M.G. & James, K. (1984) Asbestos-activated peritoneal macrophages release a factor(s) which inhibits lymphocyte mitogenesis. *Environ. Res.*, 35, 104–114

Donaldson, K., Bolton, R.E. & Brown, D.M. (1988a) Inflammatory cell recruitment as a measure of mineral dust toxicity. In: Dodgson, J., McCallum, R.I., Bailey, M.R. & Fisher, D.R., eds, *Inhaled Particle Conference VI*, New York, Pergamon Press, pp. 299–305

Donaldson, K. Bolton, R.E., Jones, A., Brown, G.M., Robertson, M.D., Slight, J. Cowie, H. & Davis, J.M.G. (1988b) Kinetics of the bronchoalveolar leukocyte response in rats during exposure to equal airborne mass concentration of quartz, chrysotile asbestos or titanium dioxide. *Thorax*, 43, 159–162

Donaldson, K., Brown, G.M., Brown, D.M., Bolton, R.E. & Davis, J.M.G. (1989) The inflammation generating potential of long and short fibre amosite asbestos samples. *Br. J. Ind. Med.*, 46, 271–276

Donaldson, K., Brown, G.M., Brown, D.M., Slight, J., Robertson, M.D. & Davis, J.M.G. (1990) Impaired chemotactic responses of bronchoalveolar leukocytes in experimental pneumoconiosis. *J. Pathol.*, 160, 63–69

Donaldson, K., Addison, J., Miller, B.G., Cullen, R.T. & Davis, J.M.G. (1994) Use of the short-term inflammatory response in the mouse peritoneal cavity to assess the biological activity of leached viteous fibers. *Environ. Health Perspectives*, 102 (Suppl. 5), 159–162

Driscoll, K.E., Deyo, L.C., Carter, J.M., Hassenbein, D.G. & Oberdorster, G. (1996) Development and implementation of an in vitro model for characterising mutation in the HPRT gene of rat lung epithelial cells. Paper presented at the 5th International Inhalation Symposium, Hannover. February 1995.

Folkesson, H.G., Leanderson, P., Westrom, B.R. & Tagesson, C. (1993) Increased lung to blood passage of polyethylene glycols after instillation of ferritin and asbestos fibres in the rat. *Eur. Respir. J.*, 6, 96–101

Gelzleichter, T.R., Bermudez, E., Mangum, J.B., Wong, B.A., Everitt, J.I. & Moss, O.R. (1996) Quantitation of pulmonary and pleural responses in Fischer 344 rats following short-term inhalation of a synthetic vitreous fibers. II. Pathobiologic responses. *Fundam. Appl. Toxicol.*, 30, 39–46

Goodglick, L.A. & Kane, A.B. (1990) The role of fibre length incrocidolite asbestos toxicity in vitro and in vivo. In: *VII International Pneumoconiosis Conference* Part 1 (DHHS (NIOSH) Publication No 90-108), pp. 163–169

Hamilton, J.A. (1980) Macrophage stimulation and the inflammatory response to asbestos. *Environ. Health Perspectives*, 34, 69–74

Hannant, D., Donaldson, K. & Bolton, R.E. (1985) Immunomodulatory effects of mineral dust. 1. Effects of intraperitoneal dust inoculatoin on splenic lymphocyte function and humoral immune responses in vivo. *J. Clin. Lab. Immunol.*, 16, 81–85

Janssen, Y.M.W., Heintz, N.H. & Mossman, B.T. (1995) Induction of c-fos and c-jun proto-oncogene expression by asbestos is ameliorated by N-acetyl-L-cysteine in mesothelial cells. *Cancer Res.*, 55, 2085–2089

Kamp, D.W., Gracefa, P., Pryor, W.A. & Weitzman, S.A. (1992) The role of free radicals in asbestos-induced diseases. *Free Radical Biol.*, 12, 293–315

Koerten, H.K., Brederoo, P., Ginsel, L.A. & Daems, W.T. (1986) The endocytosis of asbestos by mouse peritoneal macrophages and its long-term effect on iron accumulation. *Eur. J. Cell Biol.*, 40, 25–36

Lee, K.P., Kelly, D.P., O'Neal, F.O., Stadler, J.C. & Kennedy, G.L., Jr (1988) Lung response to ultrafine Kevlar aramid synthetic fibrils following 2-year inhalation exposure in rats. *Fundam. Appl. Toxicol.*, 11, 1–20

Li, X.Y., Brown, G.M., Lamb, D. & Donaldson, K. (1991) Increased production of plasminogen activator inhibitor in vitro by pleural leukocytes from rats intratracheally instilled with crocidolite asbestos. *Environ. Res.*, 55, 135–144

Li, X.Y., Lamb, D. & Donaldson, K. (1992) Intratracheal injection of crocidolite asbestos depresses the secretion of TNF by pleural leukocytes in vitro. *Exp. Lung Res.*, 18, 359–372

Li, X.Y., Lamb, D. & Donaldson, K. (1993a) The production of TNFα and IL-1-like activity by bronchoalveolar leucocytes after intratracheal instillation of crocidolite asbestos. *Int. J. Exp. Pathol.*, 74, 403–410

Li, X.Y., Lamb, D. & Donaldson, K. (1993b) Interleukin 1 production by rat pleural leukocytes in culture after intra-tracheal instillation of crocidolite asbestos. *Br. J. Ind. Med.*, 50, 90–94

Li, X.Y., Brown, G.M., Lamb, D. & Donaldson, K. (1993c) Reactive pleural inflammation caused by intra-tracheal instillation of killed microbes. *Eur. Respir. J.*, 6, 27–34

Li, X.Y., Lamb, D. & Donaldson, K. (1994) Mesothelial cell injury caused by pleural leukocytes from rats treated with intratracheal instillation of crocidolite asbetos or Corynebacterium parvum. *Environ. Res.*, 64, 181–191

McConnell, E.E., Hall, L. & Adkins, B. (1991) Studies on the chronic toxicity (inhalation) of wollastonite in Fischer 344 rats. *Inhal. Toxicol.*, 3, 323–337

McFadden, D., Wright, J.. Wiggs, B. & Churg, A. (1986) Cigarette smoke increases the penetration of asbestos fibers into airway walls. *Am. J. Pathol.*, 123, 95–99

Moalli, P.A., MacDonald, J.L., Goodlick, L.A. & Kane, A.B. (1987) Acute injury and regeneration of the mesothelium in response to asbestos fibers. *Am. J. Pathol.*, 128, 426–445

Morgan, A., Collier, C.G. & Kellington, J.P. (1993) Distribution of glass fibres in the peritoneal cavity of the rat following administration by intraperitoneal injection. *J. Toxicol. Environ. Health*, 38, 245–256

Mossman, B.T., Marsh, J.P., Sesko, A., Shatos, M., Doherty, J., Petruska, J., Adler, K.B., Hemenway, D., Mickey, R., Vacek, P. & Kagan, E. (1990) Inhibition of lung injury, inflammation and interstitial pulmonary fibrosis by polyethylene glyco-conjugated catalase in a rapid inhalation model of asbestosis. *Am. Rev. Respir. Dis.*, 141, 1266–1271

Oberdörster, G., Ferin, J., Marcello, N.C. & Meinhold, S.H. (1983) Effect of intrabronchially instilled amosite on lavageable lung and pleural cells. *Environ. Health Perspectives*, 51, 41–48

Pinkerton, K.E., Plopper, C.G., Mercer, R.R., Roggli, V.L., Patra, A.L., Brody, A.R. & Crapo, J.D. (1986) Airway branching patterns influence asbestos fiber location and the extent of tissue injury in the pulmonary parenchyma. *Lab. Invest.*, 55, 688–695

Pinkerton, K.E., Brody, A.R., Miller, F.J. & Crapo, J.D. (1989) Exposure to low levels of ozone results in enhanced pulmonary retention of inhaled asbestos fibres. *Am. Rev. Respir. Dis.*, 140, 1075–1081

Quinlan, T.R., Marsh, J.P., Janssen, M.W., Leslie, K.O., Hemenway, D., Vacek, P. & Mossman, B.T. (1994) Dose-responsive increases in pulmonary fibrosis after inhalation of asbestos. *Am. J. Resp. Crit. Care Med.*, 150, 200–206

Rahman, I., Li, X.Y., Donaldson, K. & MacNee, W. (1996) Glutathione homeostasis in alveolar epithelial cells in vitro and lung in vivo under oxidative stress. *Am. Physiol. J.*, 269, L285–L292

Schapira, R.M., Ghio, A.J., Effros, R.M., Morrisey, J., Dawson, C.A. & Hacker, A.D. (1994) Hydroxyl radicals are formed in the rat lung after asbestos instillation in vivo. *Am. J. Respir. Cell Mol. Biol.*, 10, 573–579

Szymaniec, S., Donaldson, K., Brown, D.M., Chladzynska, M., Jankowska, E. & Polikowska, H. (1990) Antibody producing cells in the spleens of mice treated with pathogenic mineral dust. *Br. J. Ind. Med.*, 46, 724–728

Warheit, D.B. (1993) Assessment of pulmonary toxicity following short-term exposure to inhaled fibrous materials. In: Warheit, D., ed., *Fibre Toxicology*, New York, Academic Press, pp. 207–228

Warheit, D.B., Chang, L.Y., Hill, L.H., Hook, G.E.R., Crapo, J.D. & Brody, A.R. (1984) Pulmonary macrophage accumulation and asbestos lesions at sites of fiber depositons. *Am. Rev. Respir. Dis.*, 129, 301–310

Warheit, D.B., Hartsky, M.A., Frame, S.R. & Buttereick, C.J. (1994) Comparison of pulmonary effects in rats exposed to size-separated preparations of para-aramid or chrysotile asbestos fibres after 2 week inhalation exposures. In: Davis, J.M.G. & Jaurand, M.-C., *Cellular and Molecular Effects of Synthetic Dusts and Fibres*, Berlin, Springer-Verlag, pp. 285–298

Corresponding author
K. Donaldson
Department of Biological Sciences, Napier University, 10 Colinton Road, Edinburgh, EH10 5DT, United Kingdom

Evaluation and use of animal models to assess mechanisms of fibre carcinogenicity

G. Oberdörster

Introduction

There is increasing concern about the potential health risks associated with respirable fibres. In particular, there is controversy regarding the carcinogenic potential of man-made vitreous fibres (MMVF). Discussion centres around the interpretation, usefulness and appropriateness of results of the animal tests on which the classification of the carcinogenicity to humans of airborne fibres is based. Advocates of specific in-vivo testing models argue that other methods do not meet scientific criteria and are unsound due to their use of irrelevant exposure routes, inadequate dosing and because of other design weaknesses.

Animal models are used to evaluate several aspects of health effects of fibres. These include the assessment of pulmonary toxicity and carcinogenicity, which is most important for newly developed fibres, and the investigation of mechanisms of toxicity and carcinogenicity. Furthermore, the biopersistence of a fibre deposited in the lung can be evaluated in animal studies, and the results of animal studies are potentially useful for purposes of risk assessment, including hazard identification and risk characterization. This chapter will deal mainly with the use of animal models for evaluation of carcinogenicity with consideration of underlying mechanisms. Methods vary significantly, and it is necessary to evaluate critically currently available methodologies for assessing the potential carcinogenic effects of fibres. In view of these different methodologies it often becomes very difficult to compare results from one study with those of others. It would, therefore, be desirable to establish standardized methods for the assessment of fibre carcinogenicity based on accepted scientific principles; results from such standardized methods could be used with greater confidence by the scientific and regulatory communities for the purposes of carcinogenicity classification of fibres and risk assessment. This assessment should also include consideration of specific mechanisms that may be operating under given experimental conditions.

Fundamentally different approaches have been used for the evaluation of the carcinogenicity of fibres; these include inhalation, intratracheal instillation and intracavitary injection into the pleural and peritoneal space. Advantages and disadvantages of each specific method will be examined below. In addition, some basic concepts, principles and questions related to mechanisms of fibre carcinogenicity will be reviewed, since these are important for the interpretation of test results.

Evaluation of animal models

Several animal models have been used to evaluate the carcinogenic potential of fibres. These include inhalation, intratracheal instillation and intracavitary injection, mostly performed as intraperitoneal injection (often referred to as i.p.). Table 1 compares these methodologies with respect to their advantages/limitations for specific testing parameters and objectives, including hazard identification, risk characterization and costs. A plus sign indicates that a specific method is suitable to achieve the objective, and a minus sign indicates the contrary. With regard to costs, the highest costs are incurred with the labour- and time-intensive inhalation studies, and this is indicated by a minus sign. Some of the points made in this table are discussed further in the sections *Inhalation*, *Intratracheal instillation* and *Intracavitary instillation*. However, because of the importance and relevance of fibre dose levels in carcinogenicity testing with different methods, the issue of the maximum tolerated dose (MTD) is discussed briefly first. More recently, the term MTD has also been used to mean the 'minimally toxic dose' (Bucher, 1994; Morrow *et al.*, 1996); however, the basic concept is the same, that is, the highest dose level in a multi-dose chronic bioassay should be at the MTD.

The MTD in carcinogenicity assays

As indicated in Table 1, it is not clear at present whether tumours induced by natural or man-made fibres occur in any of the three test methods at levels below the MTD. Problems associated with the use of high exposures in chronic toxicity and carcinogenicity testing have been long recognized – excessively high dose levels of toxicants can lead to effects (including tumours secondary to toxicity) that are irrelevant for the expected lower human exposures. For example, it has become increasingly clear that lung tumours in chronic rat inhalation studies with non-fibrous particles may be causally related to chronic alveolar inflammation induced by high particle burdens. Figure 1 illustrates this by outlining suggested pathogenic sequences of effects of inhaled non-fibrous particles as they are observed in rats. Major hallmarks in this sequence are acute and chronic inflammation, mutational events and epithelial cell proliferation, and these may eventually result in tumour formation. Results of a recent 13-week inhalation study with three different concentrations of carbon black (Oberdörster et al., 1995) showed a dose-dependent pulmonary inflammatory response. There was also a dose-dependent increased mutation frequency (hprt gene mutations) of alveolar epithelial cells after 13 weeks of exposure to carbon black, which paralleled the responses of the inflammatory cell influx – only those groups that showed significant alveolar inflammation also showed increased mutation frequencies (Driscoll et al., 1996). Driscoll (1995) had previously demonstrated that increased hprt mutation frequencies of lung epithelial cells co-cultured with pulmonary inflammatory cells could be diminished by antioxidants. The authors suggested that reactive oxygen species (ROS) released by inflammatory cells may be responsible for the increased mutation frequencies. Weitzman and Stossel (1981) had reached the same conclusion in an earlier study; they demonstrated increased mutation responses in *Salmonella typhimurium* after incubation with human monocytes and neutrophils; the mutation response could be diminished when neutrophils were used that were unable to generate ROS. To avoid these and other indirect effects on tumour induction, the use of an MTD in chronic carcinogenicity assays has been proposed.

The MTD should be the highest dose level in chronic studies. Sontag et al. (1976) defined the MTD as 'the highest dose of a test agent during the chronic study that can be predicted not to alter

Table 1. Comparison of animal models for assessing chronic toxicity and carcinogenicity of fibres

	Inhalation	Intratracheal instillation	Intracavitary instillation
Physiological exposure route	+	(+)	–
Physiological lung defences operable	+	(+)	–
Administration of human respirable fibres	(+)	+	+
Determination of pulmonary biopersistence	+	(+)	–
Evaluation of pulmonary toxicity	+	+	–
Assessing mechanisms for lung tumour induction	+	+	–
Assessing mechanisms for mesothelioma induction	+	+	+
Tumours at and above MTD	+	+	+
Tumours below MTD with asbestos (known human carcinogen)	?	?	?
Hazard identification (provided MTD is considered)	+	+	+
Risk characterization	+	–	–
Costs	–	+	+

MTD, maximum tolerated dose.

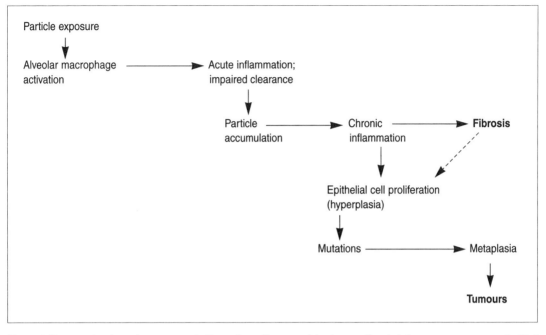

Figure 1. Suggested pathogenic sequence of effects of non-fibrous particles in rats. Chronic inflammatory events are of critical importance. Impaired lung clearance as an important early event is observed in particle overload conditions with particles of low toxicity (e.g. TiO_2) but also at lower doses (non-overload) with cytotoxic particles (e.g. crystalline SiO_2). The suggested pathogenic sequence for both types of particles appears to be the same. Fibre-induced impaired lung clearance also occurs at lung burdens lower than for non-fibrous low-toxicity particles (Yu et al., 1994; Warheit et al., 1995).

the animal's normal longevity from effects other than carcinogenicity'. The emphasis here is on 'predicted', indicating that at best the MTD can be an estimated maximum tolerated dose. It was further suggested that the MTD could be estimated from a subchronic study as a dose level causing a decrease in body weight gain of approximately 10%.

Approaches to define the MTD have been recommended by various agencies which, to a varying degree, also suggest significant histopathological changes to be used as indicators, as well as changes in toxicokinetics and metabolic parameters (IARC, 1980; OSHA, 1980; OECD, 1981; US EPA, 1982; NTP, 1984; ILSI, 1984; OSTP, 1985). For example, a nonlinear tumour response may be caused by a metabolic overload at higher doses, and this is potentially an indicator of the MTD, but only if it can be shown that a mechanism of action is operating that does not operate at the lower dose levels or in humans at anticipated human exposure levels (Haseman, 1985). The present standard NCI/NTP design for a chronic carcinogenicity study requires a life-long exposure at three dose levels in rodents, with exposures for the highest dose being at the MTD; subsequent lower doses may be half and a quarter of the MTD (Haseman, 1985). Participants of a recent workshop – devoted specifically to the MTD in particle inhalation studies – agreed that the MTD should not be exceeded in chronic carcinogenicity studies (Morrow et al., 1996).

With specific reference to inhalation studies, Lewis et al. (1989), in an NTP-sponsored workshop, discussed maximal aerosol exposure concentrations in inhalation studies. The emphasis of this workshop was on highly insoluble particles of low toxicity, formerly referred to as 'nuisance' particles. Two of several recommendations from the workshop were that (i) testing should not be performed at the highest technologically feasible concentration, but rather the highest inhaled concentration should lead to a minimum of interference with lung defence mechanisms, such as lung clearance; in addition, the two lower of the recommended concentrations should not lead to interference of normal clearance and particle accumulation in the lung; (ii) the residence of test material in the lung as a function of time should be determined; this very important parameter will readily show whether accumulation in the lung at different exposure concentrations occurred with linear or nonlinear kinetics. Adherence to these recommendations would allow the identification of any potential high-particle-load-related deviation from the normal accumulation kinetics.

In another effort addressing the issue of MTD, specifically for chronic inhalation studies, Muhle *et al.* (1990) introduced the term maximal functionally tolerated dose (MFTD) for such studies. Muhle *et al.* (1990) arbitrarily defined this dose as a lung burden causing a two- to fourfold prolongation of particle clearance based on their results with a number of particle inhalation studies in rats. This suggestion was for non-fibrous particles of low toxicity and low solubility; however, this concept of using perturbations of lung clearance function as an indicator of an MTD should also be extendible to fibres.

At a recent EPA-sponsored workshop dealing with developing guidelines for carcinogenicity testing of fibres (EPA, 1995), participants suggested that significant changes in the following endpoints should be considered for defining an MTD in rodents for fibre testing: increased lung weight; chronic pulmonary inflammation as determined from bronchoalveolar lavage (BAL) parameters; prolongation of alveolar macrophage-mediated particle clearance; increased cell proliferation; pulmonary histopathological changes; and fibre retention kinetics. Significant pulmonary toxicity indicated by these endpoints would imply that the rodent MTD was achieved. If tumours are induced both below and at this MTD, this information should obviously be used for classification of the fibre as a probable or possible human carcinogen. If, however, tumours are only induced at the MTD or at levels exceeding the MTD, then one might question how meaningful this is for extrapolation to humans, especially if it can be demonstrated that the underlying high-dose mechanism does not operate in humans who are exposed only to lower concentrations. The problem is that so far it has not been demonstrated with any of the methods considered in Table 1 that lung tumours or mesotheliomas are induced experimentally by exposure concentrations or doses of fibres – including asbestos – below MTD levels. This must be kept in mind when the different methodologies are compared in the following sections.

Inhalation studies

Inhalation is the physiological route of exposure for airborne particles (Table 1), and one might expect that chronic inhalation exposure of animals with airborne fibres will correctly identify the carcinogenic potential of a fibre, provided that the selected species is sufficiently sensitive. Usually, the rat is used as the species of choice. Both whole-body exposures and nose-only exposures have been developed, each having its role for specific applications (Phalen, 1984). Generally, whole-body exposures require more test material and will result in the dusting of the whole animal, while nose-only exposures restrict external contamination of the animal to the head and neck region. This avoids significant contamination of the laboratory environment in the post-exposure phases, which is highly important for particulate material of greater toxicity; however, there is greater stress on the animals during daily restraint for hours in the nose-only tubes and this may have an influence on toxicity outcome. As will be discussed below, hamsters appear to be specifically sensitive towards induction of mesothelioma and would, therefore, be a good choice as a second species. Generally, results from rodent inhalation studies are considered appropriate for risk characterization purposes with the caveat that extrapolation to humans has to be performed cautiously by considering important species differences. However, the inhalation route of exposure in rodents is not devoid of problems; specifically, the issues of fibre respirability and MTD deserve further discussion.

An important distinction exists between 'inhalability' and 'respirability' of inhaled particles. As defined by ACGIH (1994), inhalable particles are those that are hazardous when deposited anywhere in the respiratory tract, whereas respirable particles are hazardous when deposited in the gas exchange region. In addition, a thoracic particulate mass is defined for those materials that are hazardous when deposited anywhere within the lung airways (conducting airways) and the gas exchange region. For spherical particles, curves defining the three categories for humans are well described (ACGIH, 1994). Studies in rats (Raabe et al., 1977) as well as mathematical deposition models (e.g. Yeh & Schum, 1980) also permit a very good description of deposition efficiencies of inhaled spherical particles in this species. For example, spherical particles larger than c. 5 µm in diameter are not respirable for rats, whereas they are well respirable for humans. This example highlights the importance of respiratory tract dosimetry, and it is necessary to recognize these differences when interpreting and extrapolating results found in rats. In addition, it needs to be considered that rodents are obligatory nose-breathers whereas humans are most often mixed oro-nasal breathers. In humans breathing only orally (the worst case), significantly more inhaled non-fibrous particles penetrate to the alveolar region.

Respiratory tract dosimetry becomes even more complex when inhaled fibrous particles are considered. Several factors play a role with respect to adverse effects and respiratory tract dosimetry. One is, as discussed below, that humans develop lung tumours both in the conducting airways and in peripheral regions, with conducting airway tumours probably being the major portion. In rats, in contrast, lung tumours after inhalation of fibres are only induced in the peripheral region of the lung. This implies that respirable fibres are most important for the rat, whereas for humans, fibres conforming to the thoracic particulate definition will also have to be considered, and this includes more than the respirable fraction. Secondly, respirable fibres for humans and rats can represent very different fractions, which (as mentioned above) becomes even more evident when human mouth-breathing is considered. For example, Figs 2 and 3 show the predicted alveolar deposition efficiencies (respirable fibres) for fibres of different aerodynamic diameters and different aspect ratios (β) between rats and mouth-breathing humans. The fibre deposition curves are contrasted with those for spherical particles. It is obvious that respirability for rats is very different compared to respirability in humans. Predicted deposition curves for hamsters are very similar to those for the rat. V. Timbrell's work has suggested that the maximal diameter of human respirable asbestos fibres is 3.5 µm (Timbrell, 1965); however, in more recent studies asbestos fibres with diameters of up to 4.1 µm were detected in human lung tissue (Timbrell, 1982), and this agrees quite well with predictions from Fig. 2. Figure 2 uses the aerodynamic diameter of fibres, which is derived from their geometric diameter, their length and their specific density. Since the density of asbestos fibres is 2.4–3.3 g/cm^3, the aerodynamic diameter of asbestos is greater than that of a fibre with unit density. Figure 4 shows the relationship between geometric and aerodynamic diameters of fibres of different aspect ratios with a specific density of 2.7 g/cm^3 (i.e. refractory ceramic fibres) for random orientation of fibres. This figure can be used for converting aerodynamic diameters in Fig. 2 to geometric diameters.

The differences in fibre respirability between humans and laboratory rodents raises the following questions with respect to evaluating carcinogenicity of fibres and underlying mechanisms by inhalation: (i) Should fibre samples for these studies be prepared so that they are rodent-respirable or should they represent a human respirable sample? (ii) Since longer human respirable fibres may have significant chronic adverse effects, will these be testable by rat inhalation studies? For example, a 28 µm fibre of 1.4 µm diameter (specific density = 2.7) has an aerodynamic diameter of 3 µm (Fig. 4); this fibre is not respirable by the rat, but more than 20% of inhaled fibres of this size are deposited in the human alveolar region. Because of the importance of fibre respirability for the appropriate dosing of the lung in a chronic inhalation bioassay, the bivariate size distribution and numbers of the retained fibres in the lung need to be determined. It must be assured that enough fibres of critical dimension were retained in the lungs of the experimental animals. Participants at the recent EPA-sponsored workshop to develop fibre testing guidelines realized this problem and suggested that, if necessary, the fibre aerosol sample can be enriched with longer fibres (EPA, 1995).

The other issue requiring additional discussion with respect to fibre carcinogenicity testing by inhalation relates to the MTD. As discussed in the preceding section, endpoints to be considered for defining the MTD include lung weight, inflammation, impaired particle clearance function, cell proliferation and histopathology. A key question, however, is whether chronic inhalation of any fibre type by rats will result in lung tumours at exposure levels below the MTD. This has not been demonstrated yet since, in the past, either the respective

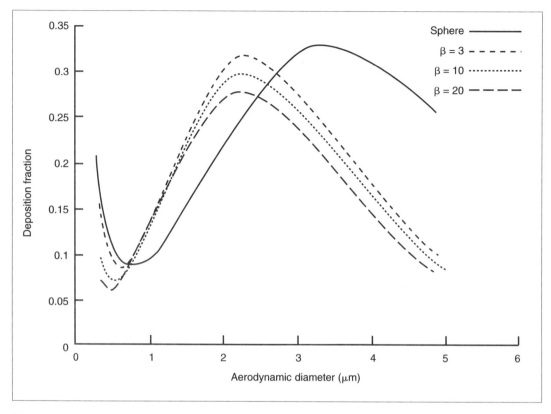

Figure 2. Model prediction of alveolar deposition of inhaled fibrous and non-fibrous particles in humans (mouth-breathing at rest) (β = aspect ratio of fibre). (Provided by C.P. Yu, personal communication.)

endpoints listed above were not determined in long-term studies or exposures were at levels that are known to induce significant toxicity.

To validate the inhalation model, a multidose inhalation study in rats and hamsters with asbestos – a known human carcinogen – is needed with the highest exposure level being at a rigorously defined MTD. If concurrent lower exposure concentrations in this multidose study do not result in the induction of lung tumours, or mesotheliomas, then one or more of the following scenarios is indicated: the mechanisms of tumour induction operating at MTD levels in rodents are also operating in humans exposed to lower concentrations; asbestos-induced tumours in humans are the result of exposure at MTD and higher levels; the rodent inhalation model is inadequate (e.g. due to differences in respirability); or our concept of defining an MTD and performing the study by using the MTD as a maximal exposure level is inappropriate.

One obvious difficulty is how to decide when a response observed in a target organ truly indicates an MTD rather than reflecting a normal host defence response. For example, when is an increase in activation and recruitment of alveolar macrophages and pleural macrophages, or an increase in cell proliferation (an inflammatory response to a high dose), an indication of the MTD being exceeded and when is it a 'physiological' response to the agent even at doses lower than the MTD? High priority should be given to resolving this issue which is crucial for establishing appropriate criteria for the chronic inhalation model in rodents with the aim to assess fibre carcinogenicity. Without this, the relevance of this model is questionable.

Intratracheal instillation

Another mode of administration to the respiratory tract is via intratracheal instillation. Intratracheal instillation of particles to be tested has been

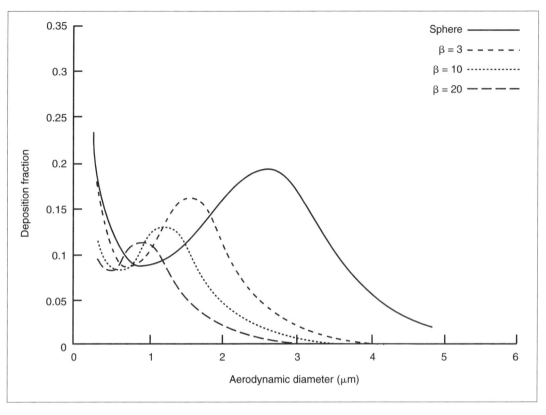

Figure 3. Model prediction of alveolar deposition of inhaled fibrous and non-fibrous particles in rats (β= aspect ratio of fibre). (Provided by C.P. Yu, personal communication.)

performed frequently in rats, hamsters and mice. In contrast to inhalation, the particles are delivered within a fraction of a second at doses that are unevenly distributed within different lung lobes, forming areas of greater deposition which can give rise to local acute inflammatory responses (bolus effect). Moreover, clumping of fibres can occur at higher instilled doses, especially when long fibres are present, which adds further to the formation of high local doses. Thus, this method is considered nonphysiological and may not be suitable for purposes of risk characterization. However, the method could well be considered for comparative studies to establish dose–response relationships for the purpose of hazard identification and toxicity ranking of different fibre types. An advantage is that human respirable fibres can be delivered to the rodent lung at predetermined doses. In addition, instilled fibres are also subjected to mechanisms affecting their biopersistence that may be the same as those acting on inhaled fibres provided that low instilled doses have been used. Repeated instillations at weekly intervals for up to 20 weeks – or even longer – can be performed and the animals kept for the rest of their lives to evaluate carcinogenic responses (e.g. Pott *et al.*, 1987). On the other hand, the physiological function of the mucociliary escalator may be adversely affected by high instilled doses. While recognizing the limitations of the intratracheal instillation studies, they may be very useful for answering specific questions regarding comparative pulmonary toxicity and retention characteristics of particles, but their results are not suited for the risk characterization of carcinogenicity.

While human respirable fibres can generally be delivered to rodent lungs by this method, the question of the MTD needs to be considered. More likely than not, instillation of a bolus of fibres will result in inflammation, possibly cell proliferative

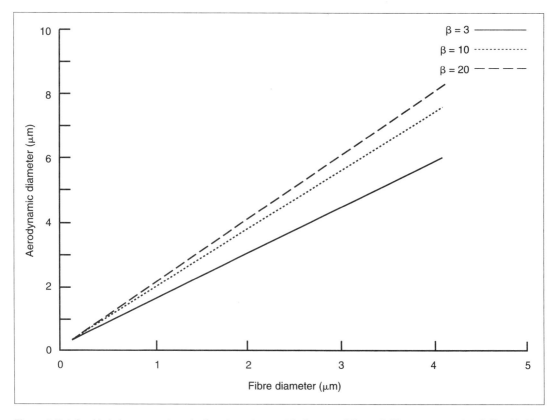

Figure 4. Relationship between aerodynamic diameter and geometric diameter of fibres of different aspect ratios, β. (Provided by C.P. Yu, personal communication.)

responses and other indicators of the MTD. Further, data on the persistence of acute toxicity signs following instillation are needed. As with the other methods of administration, the MTD needs careful attention, and as with inhalation, a multidose asbestos study would be needed to validate the intratracheal instillation method to assess the pulmonary carcinogenicity of fibres. Provided that the MTD question can be appropriately addressed, intratracheal injection could be a useful tool to identify a carcinogenic potential.

Intracavitary injections

The disadvantage of nonphysiological administration associated with the intratracheal instillation method is more obvious with routes of administration that circumvent the normal route of entry into the respiratory tract via conducting airways, that is, injections and implantations into the pleural cavity and peritoneal cavity. Both cavities are lined with mesothelial cells and, therefore, a carcinogenic response is characterized by the development of mesotheliomas when fibrous particles are directly administered in sufficiently high doses. The method of administration has the advantage of being less labour-intensive and time-consuming than the inhalation mode, and repeated injections over several months (weekly intervals) can be performed. Intracavitary injections have been used to demonstrate the carcinogenicity of asbestos and man-made fibres (Stanton et al., 1977; Pott et al., 1987).

Advocates of this method view it as an excellent and easy-to-perform bioassay to determine both the carcinogenic potential and carcinogenic potency of a fibre, the results not only being relevant for the development of mesothelioma but also relevant for indicating a carcinogenic risk for lung tumours. The fact that intrapleural and intraperitoneal injections of different asbestos fibres result in a tumorigenic response that has also been found in exposed humans either as mesothelioma, lung cancer or both, is cited as convincing evidence for the relevance and appropriateness of this mode of application. Indeed, IARC accepts findings of carcinogenicity from this route of administration, and in its Preamble on the Evaluation of Carcinogenic Risks to Humans (e.g. IARC, 1993) makes the very general statement '... in the absence of adequate data on humans, it is biologically plausible and prudent to regard all agents and mixtures for which there is sufficient evidence of carcinogenicity in experimental animals as if they presented a carcinogenic risk to humans'.

Critics of the intracavitary routes of administration argue that, in addition to being highly non-physiological, this method also bypasses all defence mechanisms of the respiratory tract, and results in massive local doses with significant inflammatory responses (i.e. exceeding the MTD) that would not be relevant for human exposure; it could at best be used for identifying a hazard to mesothelial cells but not to cells of the conducting airways and alveolar region, since the cell types are very different between the two regions and it cannot be assumed or expected that responses should be the same.

Differences to inhalation are, indeed, quite obvious. Inhalation of fibrous particles will lead to a very slow build-up in the lung only; defences such as different local clearance mechanisms (including alveolar macrophages, other inflammatory cells, mucociliary escalator and dissolution processes) are active in the lung to remove fibres after deposition by inhalation (see Fig. 5). Migration to the pleura from the alveolar space occurs for a limited number of fibres, and this translocation process appears to be different for different types of fibres. Fibres reaching the pleural cavity are potentially cleared by local lymphatic pathways. This clearance mechanism is also available for fibres administered directly into the pleural or peritoneal cavity. However, large intracavitary injected doses even of short fibres and lower doses of large non-fibrous particles can readily overwhelm and block this clearance pathway as shown by Goodglick (1988) and Goodglick and Kane (1990). Subsequently injected short or long fibres are persistently retained in this cavity and will cause cytotoxicity. Although this scenario following direct intracavitary injection is very different from situations where persistent fibres after inhalation migrate to pleural sites at a very low rate, intraperitoneal or intrapleural injections could be useful for identifying a potential hazard of a fibre for inducing mesothelioma, provided the dosage is appropriately considered.

Intraperitoneal rather than intrapleural injection has become the most widely used method for evaluating carcinogenicity of fibres when using intracavitary injection methods. The rat is used as the animal of choice for intraperitoneal injection studies. A draft protocol for intraperitoneal testing of fibres developed at the International Cooperative Research Programme on the Assessment of MMFs Toxicity (ICRP, 1994) suggested that doses to be used for each fibre sample should be 1×10^9, 1×10^8 and 1×10^7 fibres which should be injected intraperitoneally after being suspended in phosphate buffered saline in an injection volume of 2 mL. The maximum dose in terms of mass should not exceed 50 mg per injection, and when more than one injection is required to obtain the full dose, then weekly intervals of injections should be performed; however, the maximum overall dose in terms of mass should not be more than 250 mg of fibres.

These suggested doses for intraperitoneal injection studies require some further consideration with respect to the MTD. The peritoneal cavity consists of a large surface area of $c.$ 600 cm^2 in the rat (Rubin et al., 1988). If the recommended maximal dose for the intraperitoneal administration (i.e. 50 mg per injection and 250 mg total) was deposited in the alveolar region of a rat lung ($c.$ 4400 cm^2 surface area), this would be equivalent to 360 mg per rat lung per dose, or 1.8 g per rat lung maximally. This latter figure is more than the weight of a rat lung, a dose which is certainly extremely excessive. Such a lung burden could not be achieved by either inhalation or intratracheal instillation. High inflammatory and cell proliferative responses are certain to occur in the peritoneal cavity at these doses, as has been demonstrated already at much lower injected fibre burdens (Moalli et al., 1987; Goodglick & Kane, 1990). Similarly, intrapleural inoculation of fibres resulted in chronic inflammation, fibrosis and foreign body reactions (Fraire et al., 1994). The question is, therefore, how can peritoneal doses be selected to meet MTD criteria? This is of even greater importance considering that the peritoneal cavity is not designed to be exposed to environmental particulate compounds – it lacks specific clearance defences against even moderate doses, quite in contrast to the alveolar region of the lung.

As with inhalation and intratracheal instillation, the MTD criteria for the peritoneal cavity may include inflammatory, cell-proliferative and histopathological endpoints. Without proper consideration of non-excessive doses (which can be very different for different fibre types), data obtained from fibre-induced tumours will always be weakened by the possibility of the tumours being due to secondary events and mechanisms associated with high-dose toxicity.

Factors influencing carcinogenicity of fibres

Mechanisms for lung cancer versus mesothelioma

Figure 5 depicts a scheme of movement and potential effects of inhaled fibres after deposition in the respiratory tract. Starting with exposure to airborne fibres, fibres are deposited throughout the airways of the lower respiratory tract and are subsequently translocated to different regions of the respiratory system. The importance of different clearance pathways, of fibre dimension and fibre durability, and of inflammatory processes in the pathogenesis of chronic effects is indicated. Presumably, fibre-induced bronchogenic carcinoma in humans is induced by fibres depositing in the conducting airways or by fibres being cleared from more distal portions of the respiratory tract via the mucociliary escalator so that they can interact with bronchial and bronchiolar epithelial cells. In contrast, pleural mesothelioma is probably induced by fibres that translocated to the pleura and pleural space after deposition in the alveolar region. It is also indicated in Fig. 5 that lung tumours induced in rats after inhalation exposure to some fibrous materials are localized in the peripheral lung, possibly originating from type II alveolar epithelial cells, in contrast to the more centrally localized tumours in the human lung (bronchogenic carcinoma).

The translocation of fibres towards the pleura requires that fibres deposited in the alveolar space

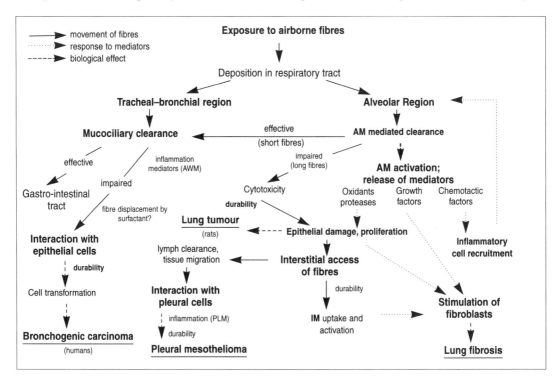

Figure 5. Hypothetical pathogenic pathways for pulmonary effects of inhaled carcinogenic fibres in the respiratory tract of humans and rats (AWM = airway macrophages; AM = alveolar macrophages; IM = Interstitial macrophages; PLM = pleural macrophages).

are able to access the pulmonary interstitium; further transport mechanisms to the pleura could involve lymphatic pathways and possibly cellular transport by interstitial macrophages. Indeed, lymphatic translocation of amosite fibres deposited in the alveolar region has been demonstrated in dogs (Oberdörster et al., 1988). Fibres reaching the pleura represent only a small fraction of the total deposited fibre burden in the alveolar region. Moreover, one would expect that it is mostly the longer fibres that reach the pleura since (i) these cannot be readily phagocytized by alveolar macrophages for mechanical clearance via the mucociliary escalator, and (ii) the long and thin fibres have a greater potential to induce mesothelioma, based on studies performed by Stanton et al. (1981) and Pott (1978) with intrapleural and intraperitoneal injections using different types of fibres. Shorter fibres, in contrast, can be effectively phagocytized and cleared by alveolar macrophages. It appears, therefore, that clearance processes in the alveolar region should favour the retention of the potentially more dangerous fibres that could then reach the pleura.

However, in a recent 5-day inhalation study in rats with refractory ceramic fibres (RCF) only short (< 5 μm), thin fibres could be detected at pleural sites over a four-week post-exposure observation period (Gelzleichter et al., 1996). Moreover, these fibres were found immediately after cessation of exposure, implying a very rapid translocation of short fibres to the pleura, which contrasts with their predicted rapid phagocytosis and clearance via alveolar macrophages. This result is puzzling and raises the question as to whether the presumably more toxic longer fibres need to be in contact with pleural cells to induce adverse effects. Using inhalation of non-fibrous particles, other investigators have also observed that pleural inflammatory responses occurred when pulmonary inflammation was induced by these particles (Henderson et al., 1995). Thus, further studies are needed to differentiate indirect pleural responses as a consequence of particle-induced pulmonary inflammation from those induced directly by the contact of fibres with pleural mesothelial cells. For example, Adamson et al. (1993) suggested that mesothelial cells may be stimulated by cytokines released from activated interstitial macrophages which then diffuse across the interstitium. This suggestion is based on in-vivo studies using long and short crocidolite; these studies detected mesothelial cell proliferation without the appearance of fibres at the pleura. Data from Owens and Grimes (1993) concerning tumour necrosis factor α (TNFα) induced proliferation of mesothelial cells in vitro support this concept.

Obviously, direct administration of fibres, both natural and man-made, at sufficiently high doses to mesothelial sites results in the induction of mesothelioma, as has been amply demonstrated by intracavitary injection studies. However, direct bolus administration to pleural or peritoneal sites of large doses is not equivalent to a slow translocation of inhaled fibres in vivo from the alveolar space to the pleura. Despite this obvious difference in dose rate, it has been suggested that intracavitary injection will readily identify a potentially carcinogenic fibre with respect to induction of mesothelioma. Furthermore, it has been argued that induction of mesothelial tumours by intracavitary injection of a fibre would also correctly identify this fibre as an inducer of lung tumours. This argument seems to be supported by observations that (i) fibres of asbestos and erionite, which are known to cause human bronchogenic carcinoma and mesothelioma, also induce mesotheliomas after intracavitary injection in rodents; and that (ii) intraperitoneal injections of non-fibrous particles do not induce mesotheliomas (Pott et al., 1987) or signs of persistent inflammatory reactions (Friemann et al., 1990). This latter observation is also consistent with the fact that non-fibrous particles have not been shown to induce lung tumours in humans.

However, such associations do not provide compelling evidence that tumour inducing mechanisms in pulmonary epithelial cells and in pleural cells are the same. A comparison of the responses of fibrous and non-fibrous particles at both sites may illustrate this. For example, chronic high-level inhalation exposure of rats to non-fibrous particles of low toxicity (such as titanium dioxide, TiO_2; carbon black; talc) results in chronic inflammation, fibrosis, hyperplasia and lung tumours, all of which have been associated with the phenomenon of lung particle overload and are thought to occur only at extremely high particulate lung burdens (Morrow, 1988; Oberdörster, 1995). The same responses in the lung have been found in chronic rat inhalation with asbestos and RCF at much lower lung burdens. With respect to mesothelial responses, as mentioned above, high doses of non-fibrous particles do not cause mesotheliomas or persistent inflammation when administered by injection to these sites in the rat, which is quite different from their response in the lung. Thus, the fact that even high doses of non-fibrous particles do not cause mesotheliomas but can cause lung tumours points to different carcinogenic mechanisms acting at these two target sites. With respect to fibres, high doses administered by injection to pleural or peritoneal sites can induce cellular reactions at these sites, and these reactions are similar to those induced in the lung by particle overload, i.e. chronic inflammation, fibrosis, local reactive hyperplasia and tumours (mesothelioma) (Friemann et al., 1990; Fraire et al., 1994; Rutten et al., 1994). Thus, whenever only high intraperitoneal doses show these effects, it has to be considered that the resulting tumours may be due to mechanisms that do not operate at lower doses.

In conclusion, a tumorigenic response observed after direct administration of a high dose of a fibre to mesothelial sites does not necessarily imply that inhalation of the fibre will cause tumours in the lung. Very importantly, the number and translocation rate of fibres that reach the pleura after inhalation versus direct injection have to be considered. In addition to different modes of administration, possible mechanistic differences for tumour induction between the two sites need to be considered. For the same reason (and because pleural responses seem to depend on the efficiency of fibres to be translocated to the pleura, on fibre durability, as well as on secondary responses elicited in the lung), a tumorigenic response of a fibre observed in the lung does not necessarily mean that this fibre also induces mesothelioma after in-vivo inhalation exposure.

Species differences in response and relevance for humans
Significant species differences in response to inhaled particles have been observed. Chronic inhalation of high concentrations of low-toxicity non-fibrous particles such as TiO_2, talc, carbon black and diesel exhaust induce pulmonary fibrosis and lung tumours in rats but not in mice (NTP, 1993; Heinrich et al., 1995). Furthermore, the inflammatory response elicited by these particles in rats is significantly lower in mice and even less in hamsters. A comparison between rats and hamsters after chronic inhalation of RCF (Mast et al., 1995a,b; McConnell et al., 1995) also showed that hamsters were very resistant with respect to the induction of lung tumours after inhalation of fibres, whereas rats developed lung tumours at the highest exposure concentration. This result confirms those with non-fibrous particles which indicate that the rat may be the most sensitive rodent species with respect to particle-induced lung tumour responses. In mice the only fibre inhalation study (chrysotile) reported induced tumours originating from the lining of the lungs, identified by the authors as adenocarcinomas and mesotheliomas (Bozelka et al., 1983).

On the other hand, chronic inhalation studies with fibres also revealed that hamsters are most sensitive in developing mesothelioma compared to rats. Whether this species-specific sensitivity is caused by a more rapid translocation of fibres in the hamster lung towards the pleura or due to a greater sensitivity of pleural mesothelial cells in hamsters is not known. Results by Rutten et al. (1994) could favour the latter explanation; these authors found that intrapleural injection of RCF and glass fibres in both species caused a more pronounced mesothelial cell proliferation in hamsters than in rats.

Further studies are needed to understand the mechanisms for species-specific responses. Such understanding would also be necessary to decide which species is most appropriate for the extrapolation of results to humans. Do human pulmonary epithelial and pleural mesothelial cells respond like those of rats or those of hamsters (i.e. is the rat or the hamster a better model for humans)? With respect to low-toxicity non-fibrous particles, we know that the rat lung tumour response is caused by particle overload-related effects. In the rat, any persistent particle at sufficiently high lung burdens appears to induce lung tumours. It has been suggested that this response is very specific to the rat and is not relevant for humans since coal miners with heavy lung burdens of particles – which would qualify as 'particle overload' in the rat – do not show an increased tumour incidence. However, it should also be pointed out that heavily dust-overloaded rats do not necessarily respond with lung tumours; a study by Lee et al. (1985) involved three concentrations of TiO_2 (10, 50 and 250 mg/m^3), and only at the highest concentration were lung tumours induced even though the middle concentration resulted in highly overloaded lungs, with a lung burden of c. 120 mg TiO_2 per rat lung or about 40–50 mg/g control lung (Lee et al., 1986). Average dust burdens in coal miners were reported as 15 g/lung (Stöber et al., 1965), which translates into c. 15 mg/g human lung, which is still below the TiO_2 lung burden found not to cause lung tumours in rats. Thus, available data are insufficient to determine whether the most sensitive rodent species for detecting particle-induced lung tumours, the rat, is more or less sensitive than humans in its response. We cannot, at present, dismiss the rat model as irrelevant for humans until there is more compelling evidence that specific mechanisms are operating in the rat only and not in humans.

Part of understanding mechanisms relates to measuring the correct fibre dose parameters. Customarily, dose levels are expressed in terms of mass. However, with respect to fibre–cell interactions, fibre surface area may be a more appropriate measure of fibre dose. Indeed, correlating retained particle mass of different non-fibrous particle types in rat lungs with observed lung tumour incidences does not show a good fit, whereas a reasonably good correlation can be obtained when lung tumour incidences are expressed as a function of the surface area of the retained particles (Oberdörster & Yu, 1990; Driscoll, 1996). These correlations show that the high pulmonary TiO_2 mass burden of the study described above (Lee et al., 1985) has only a low surface area whereas much lower lung mass burdens of carbon black or ultrafine TiO_2 particles have much larger surface areas (Driscoll, 1996). Retained particle surface area correlated well with lung tumour incidences, whereas retained mass did not. Nothing is known about the surface area of coal dust particles in miners' lungs, but it may be as low as that of the TiO_2 in the study of Lee et al. (1985) and, therefore, fail to reach a critical dose in terms of surface area despite a high mass burden.

It should be pointed out that dosimetry is not only important for non-fibrous particles. As shown by Timbrell et al. (1988), lung fibrosis in asbestos-exposed workers correlated best with surface area rather than mass or number of fibres retained in the lung. However, a major difference between non-fibrous and fibrous particles with respect to dosimetry is particle size-related effects. For non-fibrous particles, it has been observed that smaller particles have a significantly greater effect in the lung than larger ones (Ferin et al., 1992; Oberdörster et al., 1992). In contrast, longer fibres have a greater toxic and carcinogenic potential than shorter ones (see Fig. 5). These differences can be highly important as far as species differences of effects are concerned, specifically between rodents and humans; smaller particles are respirable by both rodents and humans, but larger particles are less or not at all respirable by rodents (respirable particles are defined as those particles reaching the alveolar region of the lung upon inhalation). Thus, potentially adverse effects (for humans) are more likely to be missed by inhalation in rodents if particles are large. Yet since larger non-fibrous particles are less toxic than smaller ones, this is not a concern for the evaluation of toxicity of non-fibrous particles in rodents. It may, however, be a problem for determining the effects of human respirable long fibres in rodents if these fibres are not respirable by rodents. Thus, differences in lung dosimetry may give rise to apparent species-specific effects. Analyses of fibre lung burdens with respect to numbers of fibres and bivariate size distribution in experimental studies is, therefore, mandatory. This will be further discussed below.

Biopersistence of fibres

When Stanton et al. (1977, 1981), Pott (1978) and Pott et al. (1987) published the results of their intracavitary injection/implantation studies with fibres that demonstrated the significance of fibre dimension for carcinogenicity, they also pointed out the importance of fibre durability. Since then, it has become increasingly clear that the durability of a fibre is a decisive factor for fibre toxicity and carcinogenicity – a more durable fibre will show a greater biological effect than a less durable fibre of the same length and diameter. However, other physico-chemical properties seem to play an important role as well, and the widely held view that dose, dimension and durability (the '3 D's') may be the only factors of significance for adverse fibre effects is not always supported by data. For example, Sjöstrand et al. (1995) reported most recently that microfabricated silicon dioxide (SiO_2) and TiO_2 fibres of exactly the same dimensions elicited significantly different responses when administered to cultured alveolar macrophages;

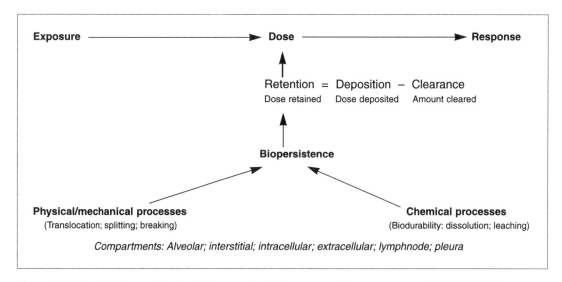

Figure 6. Relationship between different retention parameters in the exposure–dose–response paradigm for inhaled fibres.

SiO_2 fibres caused severe damage and death to the cells, whereas TiO_2 fibres induced a significantly lower response – the authors concluded that material properties played a significant role in fibre toxicity as well. As another example, chrysotile is a natural mineral fibre, the fibrogenic and carcinogenic properties of which are well recognized. In a review of the literature data, Oberdörster (1994) showed that chrysotile has a relatively low persistence in the lungs of baboons (Rendall, 1988) and rats (Coin et al., 1992) compared to other amphibole fibres; however, the biological effects of chrysotile (lung fibrosis and tumours) are no different from those of amphiboles (Wagner et al., 1974); as reviewed by Morgan (1994), leaching of magnesium may be a major factor for the rapid dissolution of chrysotile in the lung. These examples provide evidence that a sole focus on fibre durability and fibre dimension as determinants of toxicity and carcinogenicity is not always justifiable.

Nevertheless, biopersistence is one of the most important parameters for the long-term effects of fibres. In general, a more durable fibre will persist long enough in the lung to reach and interact with sensitive target cells. Figure 6 shows the relationship of biopersistence to other parameters in the exposure–dose–response paradigm for inhaled particles. Biopersistence is closely related to the retention of deposited particles; it is the result of clearance functions based on physical/mechanical processes and chemical processes in the respiratory tract. Chemical processes can be simulated *in vitro* using standardized durability assays with simulated lung fluids to determine dissolution and leaching processes. Although in-vivo and in-vitro dissolution and leaching rates are likely to be different, in-vitro durability studies can correctly identify the ranking order from highest to lowest durability among different fibre types. However, in-vitro studies cannot determine the overall biopersistence of a fibre; overall biopersistence includes specific clearance and translocation mechanisms via the mucociliary escalator and transepithelial pathways (Fig. 5). In-vivo studies in experimental animals are necessary to determine mechanical clearance rates and also dissolution and breakage rates for a comprehensive evaluation of biopersistence.

Information on fibre lengths and diameters in the lung during post-exposure periods can be used as data for calculating both in-vivo dissolution and breakage rates (Yu et al., 1995). Assuming that dissolution and breakage rates are similar between different species, including humans, and knowing that the mechanical lung clearance rate is substantially slower in humans, it may then be possible to predict the biopersistence of a fibre in the human lung. Because mechanical clearance rates for highly insoluble particles are slower in humans by a factor of 8–10 compared to rats, it can be predicted that high solubility rates of fibres have a much greater impact on lowering the overall biopersistence of fibres in human lungs than in rodent lungs. For example, the mechanical clearance rate of a given fibre in the rat lung may be equivalent to a retention half-time of 100 days and in humans to a retention half-time of 1000 days; if the dissolution rate is equivalent to a retention half-time of 100 days in both species, then the overall biopersistence, expressed as a retention half-time, would be 50 days in the rat and in humans it would be 91 days. Thus, in spite of vastly different mechanical clearance rates between the two species, the overall pulmonary retention of this fibre differs only by a factor of less than two.

Comparison of animal carcinogenicity of different fibre types

Most studies of the carcinogenic potential of fibres have used inhalation methodologies. Fewer investigators have used intracavitary injection, although a large number of tests using this method has been reported, and even fewer researchers have used intratracheal instillation. In general, the results have shown that chronic inhalation by rats or hamsters has induced lung tumours and/or mesotheliomas only after inhalation of asbestos fibres (amphiboles and chrysotile), erionite, attapulgite, silicon carbide (SiC) whiskers and K-octanate, but not after exposure to MMVF, with the exception of RCF (Lee et al., 1981; Wagner et al., 1974, 1985, 1987; Hesterberg et al., 1993; Davis et al., 1995; Mast et al., 1995a,b; McConnell et al., 1995). RCF at a high exposure concentration of 30 mg/m^3 (c. 250 fibres/cm^3) induced significant lung tumours in rats and mesothelioma in hamsters. However, results of chronic inhalation studies with asbestos were not very uniform, and negative results with respect to tumour induction have been reported (e.g. Muhle et al., 1987; Pott, 1991). Different exposure methodologies and the use of asbestos fibres of different dimensions could be the reason. For example, Davis et al. (1986) showed that 12 months inhalation of 10 mg/m^3 amosite by rats resulted in lung tumour and mesothelioma induction only when long amosite was used. Short amosite (all fibres < 5 µm in length) did not produce tumours, either by inhalation or after intraperitoneal injection.

In contrast to inhalation, intracavitary injection of the fibres listed in the preceding paragraph (including various MMVF) induced peritoneal mesotheliomas, provided the fibres were of sufficient durability (Stanton et al., 1977; Pott et al., 1987; Smith et al., 1987). Intratracheal instillation in rats resulted in the induction of lung tumours after dosing with crocidolite, chrysotile, erionite and SiC (Smith et al., 1987; Pott et al., 1987, 1991), whereas glass fibres and RCF did not in one study (Smith et al., 1987) and glass fibres were positive in another (Pott et al., 1987).

The study by Smith et al. (1987) compared crocidolite, glass fibres and RCF by chronic inhalation (2 years) in rats and hamsters and by intratracheal instillation and intraperitoneal injection in rats. With the exception of intratracheally instilled crocidolite, lung tumours were not induced by any of the fibres after either inhalation or intratracheal instillation; intraperitoneal injection induced peritoneal tumours with all three fibre types. In addition, one mesothelioma was observed in the hamsters exposed by inhalation to 12 mg RCF/m^3. Muhle et al. (1987) also observed that inhalation of crocidolite and glass fibres in rats for one year did not induce lung tumours, whereas intraperitoneal injection of both fibre types resulted in mesothelioma of the peritoneal cavity; fibre length and dose may have played a role in the negative inhalation study (Muhle et al., 1987) – 90% of the inhaled crocidolite fibres were shorter than 4.5 µm at an exposure concentration of 2.2 mg/m^3 resulting in a lung burden of c. 1 mg, whereas for the intraperitoneal injection 21% of the fibres were longer than 5 µm of which a dose of 0.5 mg was injected. Based on the available cellular surface areas in the alveolar region versus the peritoneal cavity of the rat (see *Intracavitary injections*), the peritoneal cavity was dosed more than fivefold higher than the lung in the study by Muhle et al. (1987). In addition, differences in fibre dimension and in dose rate (bolus effect after intraperitoneal injection versus slow accumulation after inhalation) as well as differences in cell types and cellular defences between the two sites (see *Mechanisms for lung cancer versus mesothelioma*) have to be considered when interpreting and comparing the results of these studies.

The importance of fibre durability for tumour outcome was also demonstrated in several studies. In the study by Muhle et al. (1987), Californian chrysotile did not show a tumorigenic response, neither after chronic inhalation nor after intraperitoneal injection. The authors attributed this to the very low biopersistence of this particular chrysotile. In another study, McConnell et al. (1991) reported that a 2-year inhalation of wollastonite (10 mg/m^3) in rats did not result in tumours or significant lung pathology, whereas Canadian chrysotile induced significant numbers of bronchoalveolar tumours. Again, the low biopersistence of wollastonite is likely to be the reason for the negative outcome of the inhalation study.

The possible impact of non-fibrous particles on tumorigenicity deserves some attention. Davis et al. (1991) reported that combined exposures of rats to amosite plus TiO_2 or SiO_2 particles or to chrysotile plus TiO_2 or SiO_2 particles increased both lung and mesothelial tumours. This is of importance when evaluating and comparing the carcinogenic potency of a pure fibre sample versus a fibre sample contaminated with a significant amount of non-fibrous particles. For example, a very high percentage (52–76%) of the RCF aerosol used in the chronic inhalation study in rats and hamsters (Mast et al., 1995a,b; McConnell et al., 1995) consisted of non-fibrous particles (Yu et al., 1994) which may have contributed mechanistically to the observed tumour response via increased alveolar macrophage activation, impaired clearance function and epithelial cell proliferation (Fig. 5). Finally, it should be mentioned that erionite is specifically potent for inducing mesothelioma as has been shown in chronic inhalation studies in rats (Wagner et al., 1985; Johnson & Wagner, 1989). Twelve months of inhalation exposure to 10 mg/m^3 induced mesothelioma in 27 out of 28 rats (Wagner et al., 1985).

For most of the studies cited here, it is not possible to evaluate the result with respect to the issue of MTD. As discussed above, increased epithelial cell mutation rates and eventual manifestation of tumours may be linked mechanistically to chronic inflammatory conditions in the lung due to very high doses of persistent particles. Thus, it is very difficult to compare the carcinogenicity of different fibre types when results are based on different methodologies (inhalation versus intracavitary injection) unless dose has been appropriately considered in the individual studies and the endpoints of toxicity have been documented. Most importantly, a better understanding of mechanisms is needed for an improved evaluation of fibre carcinogenicity. In many studies, only one dose level has been used, yet guidelines for assessing carcinogenicity of fibres in animal studies recommend that exposure concentrations or doses should cover a range from very low doses up to the MTD. There is increasing evidence and acceptance of the fact that the dose defines the mechanism, but the challenge to scientists is to reach agreement on what constitutes an unacceptably high dose.

Acknowledgements
This paper was prepared with partial support from NIEHS Grants ESO1247 and RO1 ESO4872.

References
ACGIH (1994) *American Conference of Governmental Industrial Hygienists: 1993–1994. Threshold Limit Values for Chemical Substances and Physical Agents and Biological Exposure Indices*, Cincinnati, OH

Adamson, I.Y.R., Bakowska, J. & Bowden, D.H. (1993) Mesothelial cell proliferation after instillation of long or short asbestos fibres into mouse lung. *Am. J. Pathol.*, 42, 1209–1216

Bozelka, B.E., Sestini, P., Gaumer, H.R., Hammad, Y., Heather, C.J. & Salvaggio, J.E. (1983) A murine model of asbestosis. *Am. J. Pathol.*, 112, 326–337

Bucher, J.R. (1994) Critical factors for dose selection in NTP studies. In: *The Toxicology Forum, 1994 Annual Summer Meeting, July 11–15, 1994*, Aspen, CO, The Given Institute of Pathobiology, pp. 227–236

Coin, P.G., Roggli, V.L. & Brody, A.R. (1992) Deposition, clearance, and translocation of chrysotile asbestos from peripheral and central regions of the rat lung. *Environ. Res.*, 58, 97–116

Davis, J.M.G., Addison, J., Bolton, R.E., Donaldson, K., Jones, A.D. & Smith, T. (1986) The pathogenicity of long *versus*. short fibre samples of amosite asbestos administered to rats by inhalation and intraperitoneal injection. *Br. J. Exp. Path.*, 67, 415–430

Davis, J.M.G., Jones, A.D. & Miller, B.G. (1991) Experimental studies in rats on the effects of asbestos inhalation coupled with the inhalation of titanium dioxide or quartz. *Int. J. Exp. Path.*, 72, 501–525

Davis, J.M.G., Donaldson, K., Brown, D.M., Cullen, R.T., Jones, A.D., Miller, B.G., McIntosh, C. & Searl, A. (1995) Methods of determining and predicting the pathogenicity of mineral fibres. *Respir. Med.*, 89, 727–728

Driscoll, K.E. (1995) Mutagenesis in rat lung epithelial cells after *in vivo* silica exposure or *ex vivo* exposure to inflammatory cells. *Appl. Occup. Environ. Hyg.*, 10, 1118–1125

Driscoll, K.E. (1996) The role of inflammation in the development of rat lung tumours in response to chronic particle exposure. *Inhalation Toxicol.*, 8 (Suppl.), 139–153

Driscoll, K.E., Carter, J.M., Howard, B.W., Hassenbein, D., Pepelko, W., Baggs, R. & Oberdörster, G. (1996) Pulmonary inflammatory, chemokine and mutagenic responses in rats after subchronic inhalation of carbon black. *Toxicol. Appl. Pharmacol.*, 136, 372–380

EPA (1995) *Workshop on Chronic Inhalation Toxicity and Carcinogenicity Testing of Respirable Fibrous Particles*, Chapel Hill, NC

Ferin, J., Oberdörster, G. & Penney, D. (1992) Pulmonary retention of ultrafine and fine particles in rats. *Am. J. Resp. Cell Mol. Biol.*, 6, 535–542

Fraire, A.E., Greenberg, S.D., Spjut, H.J., Roggli, V.L., Dodson, R.F., Cartwright, J., Williams, G. & Baker, S. (1994) Effects of fibrous glass on rat pleural mesothelium. Histopathologic observations. *Am. J. Respir. Crit. Care Med.*, 150, 521–527

Friemann, J., Muller, K.M. & Pott, F. (1990) Mesothelial proliferation due to asbestos and man-made fibres. Experimental studies on rat omentum. *Path. Res. Pract.*, 186, 117–123

Gelzleichter, T.R., Bermudez, E., Mangum, J.B., Wong, B.A., Moss, O.R. & Everitt, J.I (1996) Pulmonary and pleural responses in Fischer 344 rats following short-term inhalation of a synthetic vitreous fibre. II. Pathobiologic responses. *Fundam. Appl. Toxicol.*, 30, 39–46

Goodglick, L.A. (1988) *Mechanism of Asbestos Induced Cytotoxicity*, PhD Thesis, Brown University, Providence, RI

Goodglick, L.A. & Kane, A.B. (1990) Cytotoxicity of long and short crocidolite asbestos fibres *in vitro* and *in vivo*. *Cancer Res.*, 50, 5153–5163

Haseman, J.K. (1985) Issues in carcinogenicity testing: dose selection. *Fundam. Appl. Toxicol.*, 5, 66–78

Heinrich, U., Fuhst, R., Rittinghausen, S., Creutzenberg, O., Bellmann, B., Koch, W. & Levsen, K. (1995) Chronic inhalation exposure of Wistar rats and two different strains of mice to diesel engine exhaust, carbon black, and titanium dioxide. *Inhal. Toxicol.*, 7, 533–556

Henderson, R.F., Driscoll, K.E., Harkema, J.R., Lindenschmidt, R.C., Chang, I.-Y., Maples, K.R. & Barr, E.B. (1995) A comparison of the inflammatory response of the lung to inhaled versus instilled particles in F344 rats. *Fundam. Appl. Toxicol.*, 24, 183–197

Hesterberg, T.W., Miller, W.C., McConnell, E.E., Chevalier, J., Hadley, J.G., Bernstein, D.M., Thévenaz, P. & Anderson, R. (1993) Chronic inhalation toxicity of size-separated glass fibres in Fischer 344 rats. *Fundam. Appl. Toxicol.*, 20, 464–476

IARC (1980) *IARC Monographs on the Evaluation of Carcinogenic Risks to Humans*, Suppl. 2, *Long-term and Short-term Screening assays for Carcinogens: A Critical Appraisal*, Lyon, pp. 21–83

IARC (1993) *IARC Monographs on the Evaluation of Carcinogenic Risks to Humans*, Vol. 58, *Beryllium, Cadmium, Mercury and Exposures in the Glass Manufacturing Industry*, Lyon

ICRP (1994) *International Cooperative Research on the Assessment of Toxicity of Man Made Fibres. Workshop MMF*, Paris

ILSI (1984) The selection of doses in chronic toxicity/carcinogenicity studies. In: Grice, H.C., ed., *Current Issues in Toxicology*, New York, Springer Verlag, pp. 9–49

Johnson, N.F. & Wagner, J.C. (1989) Effect of erionite inhalation on the lungs of rats. In: Wehner, A.P., ed., *Biological Interaction of Inhaled Mineral Fibres and Cigarette Smoke*, Battelle Press, Columbus, OH, pp. 325–345

Lee, K.P., Barras, C.E., Griffith, F.D., Waritz, R.S. & Lapin, C.A. (1981) Comparative pulmonary responses to inhaled inorganic fibres with asbestos and fibreglass. *Environ. Res.*, 24, 167–191

Lee, K.P., Trochimowicz, H.J. & Reinhardt, C.F. (1985) Pulmonary response of rats exposed to titanium dioxide (TiO_2) by inhalation for two years. *Toxicol. Appl. Pharmacol.*, 79, 179–192

Lee, K.P., Henry, N.W., III, Trochimowicz, H.J. & Reinhardt, C.F. (1986) Pulmonary response to impaired lung clearance in rats following excessive TiO_2 dust deposition. *Environ. Res.*, 41, 144–167

Lewis, T.R., Morrow, P.E., McClellan, R.O., Raabe, O.B., Kennedy, G.R., Schartz B.A., Goche, T.J., Roycroft, J.H. & Chabra, R.S. (1989) Establishing aerosol exposure concentrations for inhalation toxicity studies. *Toxicol. Appl. Pharmacol.*, 99, 377–383

Mast, R.W., McConnell, E.E., Anderson, R., Chevalier, J., Kotin, P., Thévenaz, P., Glass, L.R., Miiller, W.C. & Hesterberg, T.W. (1995a) Studies on the chronic toxicity (inhalation) of four types of refractory ceramic fibre in male Fischer 344 rats. *Inhal. Toxicol.*, 7, 425–467

Mast, R.W., McConnell, E.E., Hesterberg, T.W., Chevalier, J., Kotin, P., Thévenaz, P., Bernstein, D.M., Glass, L.R., Miiller, W. & Anderson, R. (1995b) Multiple-dose chronic inhalation toxicity study of size-separated kaolin refractory ceramic fibre in male Fischer 344 rats. *Inhal. Toxicol.*, 7, 469–502

McConnell, E.E., Hall, L. & Adkins, B. (1991) Studies on the chronic toxicity (inhalation) of wollastonite in Fischer 344 rats. *Inhal. Toxicol.*, 3, 323–337

McConnell, E.E., Mast, R.W., Hesterberg, T.W., Chevalier, J., Kotin, P., Bernstein, D.M., Thévenaz, P., Glass, L.R. & Anderson, R. (1995) Chronic inhalation toxicity of a kaolin-based refractory ceramic fibre in Syrian golden hamsters. *Inhal. Toxicol.*, 7, 503–532

Moalli, P.A., MacDonald, J.L., Goodglick, L.A. & Kane, A.B. (1987) Acute injury and regeneration of the mesothelium in response to asbestos fibres. *Am. J. Pathol.*, 128, 426–445

Morgan, A. (1994) Letter to the Editor: The removal of fibres of chrysotile asbestos from lung. *Ann. Occup. Hyg.*, 38, 643–646

Morrow, P.E. (1988) Possible mechanisms to explain dust overloading of the lungs. *Fundam. Appl. Toxicol.*, 10, 369–384

Morrow, P.E., Haseman, J.K., Hobbs, C.H., Driscoll, K.E., Vu, V. & Oberdörster, G. (1996) Workshop overview: the maximum tolerated dose for inhalation bioassays: toxicity *vs.* overload. *Fundam. Appl. Toxicol.*, 29, 155–167

Muhle, H., Pott, F., Bellmann, B., Takenaka, S. & Ziem, U. (1987) Inhalation and injection experiments in rats to test the carcinogenicity of MMMF. *Ann. Occup. Hyg.*, 31, 755–764

Muhle, H., Bellmann, B., Creutzenberg, O., Fuhst, R., Koch, W., Mohr, U., Takenaka, S., Morrow, P.E., Kilpper, R., MacKenzie, J. & Mermelstein, R. (1990) Subchronic inhalation study of toner in rats. *Inhal. Toxicol.*, 2, 341–360

NTP (1984) Ad Hoc *Panel on Chemical Carcinogenesis Testing and Evaluation: Report of the National Toxicology Program*, Research Triangle Park, NC

NTP (1993) *Toxicology and Carcinogenesis Studies of Talc in F344/N Rats and B6C3F$_1$ Mice.* (National Toxicology Program Technical Report Series No. 421; NIH Publication No. 93-3152), Research Triangle Park, NC

OECD (1981) *Guidelines for Testing Chemicals*, Paris

OSHA (1980) Identification, classification and regulation of potential occupational carcinogens. *Fed. Regist.*, 45, 5001–5296

OSTP (1985) Chemical carcinogens: a review of the science and its associated principles. *Fed. Reg.*, II, 10371–10442

Oberdörster, G. (1994) Macrophage-associated responses to chrysotile. *Ann. Occup. Hyg.*, 38, 601–615

Oberdörster, G. (1995) Lung particle overload: implications for occupational exposures to particles. *Regul. Toxicol. Pharmacol.*, 27, 123–135

Oberdörster, G. & Yu, C.P. (1990) The carcinogenic potential of inhaled diesel exhaust: a particle effect? *J. Aerosol Sci.*, 21 (Suppl. 1), 397–401

Oberdörster, G., Morrow, P.E. & Spurny, K. (1988) Size dependent lymphatic short-term clearance of amosite fibres in the lung. *Ann. Occup. Hyg.*, 32 (Suppl. 6), 149–156

Oberdörster, G., Ferin, J., Gelein, R., Soderholm, S.C. & Finkelstein, J. (1992) Role of the alveolar macrophage in lung injury: studies with ultrafine particles. *Environ. Health Perspectives*, 97, 193–197

Oberdörster, G., Baggs, R., Gelein, R., Corson, N., Mercer, P., Nguyen, K. & Pepelko, W. (1995) Pulmonary effects of inhaled carbon black in rats. *Toxicologist*, 15, 46

Owens, M.W. & Grimes, S.R. (1993) Pleural mesothelial cell response to inflammation: tumour necrosis factor-induced mitogenesis and collagen synthesis. *Am. J. Physiol.*, 265, L382–L388

Phalen, R.F. (1984) *Inhalation Studies: Foundations and Techniques*, Boca Raton, FL, CRC Press

Pott, F. (1978) Some aspects on the dosimetry of the carcinogenic potency of asbestos and other fibrous dusts. *Staub-Reinhalt Luft*, 38, 486–489

Pott, F. (1991) Neoplastic findings in experimental asbestos studies and conclusions for fibre carcinogenesis in humans. *Ann. N.Y. Acad. Sci.*, 643, 205–218

Pott, F., Ziem, U., Reiffer, F.-J., Huth, F., Ernst, H. & Mohr, U. (1987) Carcinogenicity studies on fibres, metal compounds and some other dusts in rats. *Exp. Pathol.*, 32, 129–152

Pott, F., Roller, M., Rippe, R.M., Germann, P.-G. & Bellmann, B. (1991) Tumours by the intraperitoneal and intrapleural routes and their significance for the classification of mineral fibres. In: Brown, R.C., Hoskins, J.A. & Johnson, N.F., eds, *Mechanisms in Fibre Carcinogenesis*, New York, Plenum Press, pp. 547–565

Raabe, O.G., Yeh, H.C., Newton, G.J., Phalen, R.J. & Velasquez, D.J. (1977) Deposition of inhaled monodisperse aerosols in small rodents. In: Walton, W.H. & McGovern, B., eds, *Inhaled Particles*, Vol. 4, Part I, New York, Pergamon Press, pp. 3–21

Rendall, R.E.G. (1988) *The Retention and Clearance of Inhaled Glass Fibre and Different Varieties of Asbestos by the Lung*. MSc Thesis, Johannesburg, University of Witwatersrand

Rubin, J., Clawson, M., Planch, A. & Jones, Q. (1988) Measurements of peritoneal surface area in man and rat. *Am. J. Med. Sci.*, 295, 453–458

Rutten, A.A.J.J.L., Bermudez, E., Mangum, J.B., Wong, B.A., Moss, O.R. & Everitt, J.I. (1994) Mesothelial cell proliferation induced by intrapleural instillation of man-made fibres in rats and hamsters. *Fundam. Appl. Toxicol.*, 23, 107–116

Sjöstrand, M., Gold, J., Johansson, A., Thomsen, P., Brody, A. & Kasemo, B. (1995) Microfabricated titanium and silica fibres and surfaces: differential activation of lung macrophages *in vitro*. In: *Correlations Between* in vitro *and* in vivo *Investigations in Inhalation Toxicology, 5th International Inhalation Symposium, 20–24 February 1995, Hannover, Germany*, p. 65

Smith, D.M., Ortiz, L.W., Archuleta, R.F. & Johnson, N.F. (1987) Long-term health effects in hamsters and rats exposed chronically to man-made vitreous fibres. *Ann. Occup. Hyg.*, 31, 731–754

Sontag, J.M., Page, N.P. & Saffiotti, U. (1976) *Guidelines for Carcinogen Bioassay in Small Rodents* (DHHS Publication (NIH) 76-801), Washington DC

Stanton, M.F., Layard, M., Tegeris, A., Miller, E., May, M. & Kent, E. (1977) Carcinogenicity of fibrous glass: pleural response in the rat in relation to fibre dimension. *J. Natl Cancer Inst.*, 58, 587–603

Stanton, M.F., Layard, M., Tegeris, A., Miller, E., May, M, Morgan, E. & Smith, A. (1981) Relation of particle dimension to carcinogenicity in amphibole asbestoses and other fibrous minerals. *J. Natl Cancer Inst.*, 67, 965–975

Stöber, W., Einbrodt, H.J. & Klosterkötter, W. (1965) Quantitative studies of dust retention in animal and human lungs after chronic inhalation. In: Davies, C.N., ed., *Inhaled Particles and Vapours*, Vol. 2, Oxford, Pergamon Press, Symposium Publications Division, pp. 409–418

Timbrell, V. (1965) The inhalation of fibrous dusts. *Ann. N. Y. Acad. Sci.*, 132, 255–273

Timbrell, V. (1982) Deposition and retention of fibres in the human lung. *Ann. Occup. Hyg.*, 26, 347–369

Timbrell, V., Ashcroft, T., Goldstein, B., Heyworth, F., Meurman, L.O., Rendall, R.E.G., Reynolds, J.A., Shilkin, K.B. & Whitaker, D. (1988) Relationships between retained amphibole fibres and fibrosis in human lung tissue specimens. *Ann. Occup. Hyg.* 32 (Suppl. 1), 323–340

US EPA (1982) *Health Effects Test Guidelines*, Washington, DC, Office of Pesticides and Toxic Substances

Wagner, J.C., Berry, G., Skidmore, J.W. & Timbrell, V. (1974) The effects of the inhalation of asbestos in rats. *Br. J. Cancer*, 29, 252–269

Wagner, J.C., Skidmore, J.W., Hill, R.J. & Griffiths, D.M. (1985) Erionite exposure and mesotheliomas in rats. *Br. J. Cancer*, 51, 727–730

Wagner, J.C., Griffiths, D.M. & Munday, D.E. (1987) Experimental studies with palygorskite dusts. *Br. J. Ind. Med.*, 44, 749–753

Warheit, D.B., Driscoll, K.E., Oberdörster, G., Walker, C., Kuschner, M. & Hesterberg, T.W. (1995) Symposium overview: contemporary issues in fibre toxicology. *Fundam. Appl. Toxicol.*, 25, 171–183

Weitzman, S.A. & Stossel, T.P. (1981) Mutation caused by human phagocytes. *Science*, 212, 546–547

Yeh, H.-C. & Schum, M. (1980) Theoretical evaluation of aerosol deposition in anatomical models of mammalian lung airways. *Bull. Math. Biol.*, 42, 1–15

Yu, C. P., Zhang, L., Oberdörster, G., Mast, R.W., Glass, L.R. & Utell, M.J. (1994) Clearance of refractory ceramic fibres (RCF) from the rat lung: development of a model. *Environ. Res.*, 65, 243–253

Yu, C.P., Ding, Y.J., Oberdörster, G., Mast, R.W., Maxim, D. & Utell, M.J. (1995) In vivo dissolution of man-made vitreous fibres (MMVF) inhaled by rats. *Int. Toxicol.*, 7, (26-PD-8), Abstracts of the VII International Congress of Toxicology, Seattle, WA, July 2–6, 1995

Corresponding author
G. Oberdörster
University of Rochester, Department of Environmental Medicine, Rochester, NY 14642, USA

Mixed fibrous and non-fibrous dust exposures and interactions between agents in fibre carcinogenesis

J.M. Davis

Introduction

The pathogenicity of mineral fibres, particularly asbestos, has been examined in great detail. However, almost all of the published work has used experimental protocols of unnaturally high dose levels (orders of magnitude higher than any human exposure) and of a fibre purity as high as, if not higher than, any human exposure. The extremely high dose levels can be justified – it seems most probable that disease production requires the accumulation of a high fibre burden in any area of lung tissue and, with short-lived experimental animals, this can only be obtained by excessive exposure. However, the use of pure fibre samples deserves closer scrutiny. Pure samples do, of course, make the interpretation of experimental findings easier, but human fibre exposures associated with workplace dust will always have been a mixture of fibres and particulate materials of several kinds. Even with pure fibrous materials, dust clouds always contain non-fibrous fragments, often in much greater numbers than the fibres themselves, and there is a need to understand what effect the non-fibrous elements of a dust cloud may have in modifying fibre pathogenicity.

There are two main ways in which the inhalation of particulate material along with fibres may modify any harmful effect of the fibres themselves. The presence of a significant lung burden of particulate material along with fibres may produce a modified tissue reaction that could increase or reduce the level of disease expected from the fibres. In addition, the presence of particulate dust may modify the clearance rates of fibres from the lung and effect fibre pathogenicity in that way. Included within this category of clearance from lung tissue would be the transport of dust from the lymphatics and the pleural surface, which would have particular relevance to the production of mesotheliomas.

Mixed fibrous and non-fibrous dust exposures

A number of experimental studies have examined the pathogenicity of fibre and particulate dust mixtures. However, in most cases the protocols of these studies have not been suitable for the elucidation of the effect of dust particulates on the pathogenicity of the fibres.

Schepers (1959a,b) treated guinea-pigs, rats and rabbits by inhalation for nine months with dust produced by the sawing and abrasion of Fiberglas–plastic sheets. The animals were then maintained in a dust-free environment for a further six months. Schepers reported that the inhaled material behaved as a biologically inert substance; since neither pure Fiberglas nor pure plastic dust were used on their own, it could not be determined if the mixing of the dusts had any effect on the tissue reactions.

Wehner et al. (1978a,b) exposed hamsters by inhalation to asbestos–cement dust at two dose levels for up to 15 months. Slight dose-related increases in the incidence and severity of pulmonary fibrosis occurred with the mixed dust, but no studies were carried out with either pure cement dust or pure chrysotile asbestos at the same doses.

Davis et al. (1980) treated rats by inhalation to dusts collected in asbestos factories using either amosite or chrysotile asbestos. The amosite sample consisted of 90% pure amosite while the chrysotile sample was only 60% pure chrysotile; in each case, the remainder of the dust consisted of a mixture of the normal particulates (and non-asbestos fibres) found in the factory environment. The dose level was 10 mg/m^3 and exposure was for 12 months with full life-span follow-up. Data were available for similar exposures to pure amosite and chrysotile but it was not possible to match the fibre size distributions of the factory dusts with those from the pure dust exposures. In spite of a lower mass of asbestos in the factory dust exposures, the lung dust contents of the amosite and chrysotile at the end of 12 months was higher than with the pure asbestos exposures. The factory amosite dust produced more fibrosis than the pure amosite, while factory chrysotile produced similar levels to pure chrysotile. The factory chrysotile produced fewer malignant pulmonary tumours than pure chrysotile. Neither of the amosite samples produced any malignant pulmonary tumours.

Luchtel et al. (1989) exposed rats by intratracheal injection to a series of graphite fibre–epoxy and Fiberglas–epoxy dusts obtained by machining standard production materials by routine methods. Each animal received a single intratracheal injection of 5 mg composite dust and, as positive and negative controls, separate groups of rats received injections of either quartz or aluminium oxide. Pathological changes leading to fibrosis were graded on a four-point scale – all of the composite dusts produced changes between those found with quartz and aluminium oxide. Once again, however, no data were available regarding the effects of the fibres or the epoxy dusts on their own.

Davis et al. (1991a) compared the carcinogenicity in rats of six samples of tremolite which ranged from pure asbestiform materials with few particles to completely non-fibrous tremolite. Each animal received a single intraperitoneal injection of dust and the number of mesotheliomas produced was found to be related to the number of long fibres present; non-fibrous tremolite produced only occasional tumours, a result comparable with that found in controls. The results, however, give no information as to whether or not the presence of particulate tremolite either increased or reduced the carcinogenicity of long tremolite fibres.

Since the original work of Stanton and Wrench (1972) and Pott and Friedrichs (1972), it has been realized that long fibres are much more carcinogenic than short fibres. Davis et al. (1986) demonstrated by both inhalation and intraperitoneal injection that amosite fibres < 5 μm in length had no fibrogenic or carcinogenic potential. It is probable, therefore, that the presence of these very short fibres in an asbestos dust exposure will have the same effect as non-fibrous material of the same chemical composition.

One possible way in which high levels of short fibres or particles might effect the pathogenicity of inhaled long fibres is through the modification of clearance rates. Macrophages laden with particulate material plus a few long fibres may possibly be less motile and therefore less likely to be removed from the lung than macrophages with the same amount of long fibres alone. Some evidence on this matter has recently been produced with experimental studies of glass microfibre which included comparisons with both amosite and silicon carbide (SiC) whiskers (Davis et al., 1996). All three dusts were administered to rats by inhalation at dose levels matched for the number of fibres > 5 μm in length. Because of the necessity of milling the original glasswool to produce respirable dust clouds, however, the microfibre cloud had an order of magnitude more of short material than the other two. In spite of this, there were no significant differences in the clearance rates of long fibres from the lung for the three dust types between the end of a 12-month dusting period and 12 months later. Particulate materials or short fibres themselves can be cleared much faster from the lung than long fibres either via the bronchial tubes or to the lymphatics. This preferential removal of short material also applies to the body cavities as was demonstrated for the mouse peritoneal cavity by Goodglick and Kane (1990). In this latter study, it was found that, following intraperitoneal injection of long and short samples of crocidolite asbestos, the short material could pass through the lymphatic stomata to the regional lymph nodes while the long fibres accumulated in the stomata sites but remained in the peritoneal cavity.

Evidence that some types of particulate material can influence the pathogenicity of fibres was produced by Davis et al. (1991b) in rat inhalation studies. The effects of exposure to clouds of both amosite and chrysotile were compared with similar fibre doses to which had been added particles of either quartz or titanium dioxide (TiO_2). The asbestos dose in each case was 10 mg/m^3 for 12 months. One group of rats received in addition 10 mg/m^3 TiO_2 while another received 2 mg/m^3 quartz. Both fibre–particulate exposures caused more fibrosis and pulmonary tumours than the two asbestos samples alone. With quartz, which is both fibrogenic and carcinogenic to rats in its own right (Dagle et al., 1986; Holland et al., 1986; Muhle et al., 1991), this could have been due to a simple additive effect. However, although TiO_2 by inhalation has produced pulmonary tumours in rats at dose levels of 250 mg/m^3, no effect has been detected at 10 mg/m³ (Lee et al., 1985). The increased pathogenicity of asbestos–TiO$_2$ mixtures is therefore most likely to be due to a factor such as the modification of the clearance rate of the asbestos fibres; the study did not produce reliable clearance data to confirm this. One finding from the study of Davis et al. (1991b) that appears to result from modified clearance or transport of fibres concerns the production of mesotheliomas. Both the amosite–quartz and chrysotile–quartz mixtures produced more mesotheliomas than the two asbestos dusts alone, with the amosite figures being the highest reported in the literature for any asbestos fibre administered to rats by inhalation. Previous studies by Vincent et al. (1987) had shown in rats that quartz dust was transported through the lymphatics faster than other particulate materials. In addition, Davis (1989) had shown that, following exposure of rats to coal mine dusts with high quartz levels, lung granulomas developed on the visceral pleural surface outside the external elastic laminar of the lung and that the granulomas contained both coal and quartz particles. Since similar granulomas were not found in rats treated with coal containing very low quartz levels, it appeared likely that the quartz was not only penetrating the pleural surface itself but in some ways facilitating the penetration of the coal particles.

In the mixed asbestos–particle dust studies of Davis et al. (1991b), similar granulomas were found on the visceral pleura of rats, and these granulomas contained both quartz particles and quite long asbestos fibres. It is logical to suggest that the presence of quartz may have aided the penetration of asbestos fibres through the lymphatics to the pleural surface and that this increased penetration of asbestos was related to the increased production of mesotheliomas. Furthermore, quartz is a ubiquitous dust in the environment and its inhalation along with asbestos could have an effect on mesothelioma production in humans. One study may report an example where this has occurred. Finkelstein (1983) reported that workers in an asbestos cement factory had a high number of mesotheliomas, which was unusual especially since only chrysotile was processed. The asbestos cement mixture was, however, reported to contain 'silica'. Many pathologists claim that most human mesotheliomas develop on the parietal rather than on the visceral pleural surfaces and Jones (1987) has suggested that this is due to asbestos fibres being 'milked around the subpleural lymphatics until they reach a point of stasis'. Recent studies by Rey et al. (1994) examined the dust content of anthracotic areas of the parietal pleura in humans, some of whom were known to have been exposed to asbestos. Asbestos fibres were present among the dust particles, particularly in those subjects with a record of asbestos exposure; in this latter group, fibre levels were as high as 8×10^6 fibres/g dried lung. Since the tissue samples were from anthracotic areas, a high level of particulate dust was present (and although the authors did not report on this, a significant content of quartz is possible).

In conclusion, human exposures to asbestos or other mineral fibres always occur in a mixed dust environment. Studies indicate that the presence of particles (either fibre fragments or other mineral types) may effect the biological activity of inhaled fibres. At present, the data are too scanty to enable a proper evaluation of these effects.

Fibres as carriers of chemical carcinogens

Mixtures of quartz and asbestos either in the lung parenchyma or after transfer to the pleura could act as co-carcinogens, but the carcinogenic effect of asbestos in conjunction with organic molecules has received much greater study. This particularly relates to the effects of asbestos and cigarette smoke. Selikoff et al. (1968) first demonstrated that a synergistic relationship exists between asbestos exposure and cigarette smoking for the production of bronchial carcinomas. While the risk for asbestos workers who smoked heavily was increased more than 90 times more than that of the general population, nonsmoking asbestos workers showed a relatively small increase in tumour incidence compared to nonsmokers in the general population. While this effect is quite clear, the reasons for the added potency of asbestos fibres when combined with cigarette smoke are still uncertain. One theory suggests that no true synergism is involved but that smoking so retards pulmonary clearance that the retained dose of asbestos fibres is such that carcinogenesis based on fibres alone becomes much more likely.

Unfortunately, clear evidence is difficult to obtain from humans; data so far available have often been conflicting. Luchsinger et al. (1968) found that the fastest lung clearance time of particles labelled with radioactive chromium was found among the heaviest smokers. Thomson and Pavia (1973) found no difference in clearance rates between smokers and nonsmokers. Canmer and Philipson (1971) examined the tracheobronchial clearance of tagged particles of fluorinated ethylene propylene and found that acute exposure to tobacco smoke had no effect. Lourenco et al. (1971) examined the clearance of labelled iron oxide in humans. They found some effects during the first few hours but, by 24 h, there was no significant difference between smokers and nonsmokers. While it has been suggested that cigarette smoking reduces pulmonary clearance through the impairment of ciliary action, Binns (1975) reported that ciliary impairment is transitory and the overall effect of smoking is to increase clearance. On the other hand, Cohen et al. (1979) found that smokers showed reduced clearance of magnetite particles from their lungs and Bohning et al. (1982) found a similar clearance retardation of labelled polystyrene particles from the lungs of smokers.

Evidence from animal studies appears to show more uniform evidence that cigarette smoke can reduce the pulmonary clearance of dusts. Ferin et al. (1965) found that smoking reduced the clearance of silica in rats. Park et al. (1977) reported that smoking reduced tracheal clearance as well as macrophage activity in dogs; McFadden et al. (1986) stated that smoking greatly reduced the clearance of amosite fibres in guinea-pigs. Muhle et al. (1989) found that cigarette smoke doubled the retention of crocidolite asbestos in rats but there was no similar effect on the clearance of chrysotile. Churg and Wright (1989) found that the clearance of asbestos fibres in guinea-pigs following intratracheal injection was reduced by cigarette smoke and Mauderley et al. (1989) found a similar effect of smoking on particle clearance in rats. One explanation of the difference between animal and human studies could be that, while cigarette smoke does initially have a marked effect on particle clearance from the clean lungs of animals, the lungs of human smokers adapt with time so that impairment of clearance becomes less easily demonstrated.

Since cigarette smoke contains a number of well known chemical carcinogens, particularly polycyclic aromatic hydrocarbons (PAH) which are readily adsorbed on to asbestos fibres, it has been assumed that synergism somehow relates to this adsorption. Indeed, it was originally suggested that the carcinogenicity of asbestos fibres might be entirely caused by the absorption of hydrocarbons during processing and particularly from packaging materials such as hessian bags. This was disproved by Harrington et al. (1974) who showed that asbestos carcinogenicity in experimental animals was not affected by the complete removal of adsorbed hydrocarbons by benzene.

Using a variety of in-vitro techniques, Kandaswami and O'Brien (1980), Lakowicz and Bevan (1980), Reiss et al. (1983) and Poole et al. (1983) all reported changes in the biological activity of asbestos fibres in the presence of chemical carcinogens. In some mutagenicity studies, asbestos alone had no effect while the combined effect of the two materials was much greater than that of the chemical carcinogen on its own. Similar results have been obtained from animal experimental studies. Pylev and Shabad (1973) and Shabad et al. (1974) reported work involving the intratracheal injection of Russian chrysotile asbestos into rats with and without the administration of benzo[a]pyrene. The injected chrysotile dose was 2 mg and the benzo[a]pyrene was either adsorbed onto chrysotile before injection or, in another experiment, given as a separate injection of 5 mg. No tumours developed with either chrysotile or benzo[a]pyrene alone but in both experiments using these materials together both pulmonary tumours and mesotheliomas developed. Inhalation studies using chrysotile asbestos, benzo[a]pyrene and cigarette smoke (Shabad et al., 1974) did not cause pulmonary tumours with any combination of materials, although precancerous lesions were found more frequently in animals exposed to asbestos and cigarette smoke; the effect of asbestos and benzo[a]pyrene was more marked. It should be noted that exposure to cigarette smoke administered between periods of asbestos inhalation was trivial (1.5 cigarettes/rat/month) and exposure to benzo[a]pyrene was by means of a single intratracheal injection of 1 mg/animal.

Wehner et al. (1975) and Wehner (1980) reported more detailed inhalation studies with asbestos and cigarette smoke. Exposure of hamsters to chrysotile asbestos at a respirable mass concentration of 23 mg/L resulted in severe pulmonary asbestosis by 11 months and an increase in the incidence of benign pulmonary tumours.

Concurrent exposure to cigarette smoke (for 10 minutes, three times a day, five days a week) produced no additional carcinogenic effect.

Harrison and Heath (1991) examined the co-carcinogenic effects of chrysotile asbestos and *N*-nitrosoheptamethyleneimine by giving rats intratracheal injections of chrysotile followed by subcutaneous injections of *N*-nitrosoheptamethyleneimine. While neither the chrysotile or *N*-nitrosoheptamethyleneimine alone produced significant numbers of pulmonary tumours, the combined treatment produced pulmonary carcinomas.

While the synergism between fibres and chemical carcinogens is well demonstrated, the mechanisms involved remain uncertain. This is particularly true for the combined effects of asbestos and cigarette smoke – the major site of fibre and smoke particle deposition is in the alveoli and the bronchiolo-alveolar duct regions of the lung parenchyma, where asbestos-related tumours develop in rats, but most human lung cancers occur in the larger bronchial tubes. The two carcinogens could, of course, produce their effect by arriving separately in the lung, as must happen to a large extent, but, because PAH materials adsorb readily onto asbestos fibres, it was originally assumed that coated fibres were more potent carcinogenic agents than uncoated fibres. However, Gerde *et al.* (1994) pointed out that (i) in a moist atmosphere, such as in the inhaled air where asbestos fibres and cigarette smoke would normally meet, absorption of PAH onto fibres is very poor; and (ii) once deposition occurs on the moist alveolar or bronchiolar surfaces, liberation of bound PAH from fibres is extremely fast. Nevertheless, although PAH transport from the alveolar regions to blood is almost instantaneous, transport from the surfaces of the larger bronchial tubes is much slower, and this increases the chances of a carcinogenic effect at this latter site. Gerde *et al.* (1994) demonstrated that PAH materials are normally present in the lipid phase in the mucous layer of the bronchial epithelium and are therefore located furthest from the epithelial cells. However, Gerde *et al.* (1994) also postulate that the penetration of this mucous layer by long fibres may provide lipophilic channels where the transportation of PAH to the epithelial cells is greatly increased. There are other ways in which asbestos and cigarette smoke may combine to increase carcinogenicity. Churg *et al.* (1989) demonstrated that, in human tracheal explants *in vitro*, combined treatment with asbestos and cigarette smoke caused greater uptake of fibres by epithelial cells than asbestos treatment alone. A further possibility is that synergism between asbestos and cigarette smoke does not require the two agents to be present at the same sites. Indeed, one theory for asbestos carcinogenesis is that the main effect of the fibres is to produce a large area of tissue damage involving chronic inflammation and eventual fibrosis. In these sites, inflammatory cells secrete a number of agents (including growth factors), and this causes the chronic proliferation of epithelial and other cells and creates an environment where neoplasmic transformation of cells without further external stimulus becomes very likely. Certainly, in humans, an increased incidence of lung cancer in asbestos workers is only found in populations showing significant asbestosis. In a lung with constant high levels of growth factors in the tissue fluids, significant amounts of these factors would certainly reach the lining cells of the bronchial tubes. If these cells were already initiated by cigarette smoke acting alone, the growth factors could provide the final stimulus to neoplastic change.

Fibres as carriers of endogenous DNA

While much work has examined the possibility that asbestos fibres might carry chemical carcinogens to cells, a similar concept suggests that fibres might specifically carry DNA extruded from dead or dying cells into living cells, and that this DNA could effect transformation. For many years it has been known that cell-free DNA can cause inheritable genetic changes in bacteria (Avery *et al.*, 1944). It is also recognized that certain mineral particulate substances are very effective in mediating this process in eukaryotic cells. Both inheritable genetic changes (transformation) and the mere introduction into a cell of a functional unit of exogenous DNA (transfection) are possible. Calcium apatite crystals are particularly effective in these processes and have been used in many studies (Dubes & Klinger, 1961; Graham & van der Eb, 1973). Work by Appel *et al.* (1988) has demonstrated that asbestos fibres are able to transfect mammalian cells with exogenous DNA. Originally, monkey COS-7 cells were used in these studies but, more recently, a permanent line of human mesothelial cells has been studied (Gan et al. 1993); both chrysotile and amphibole asbestos have been shown to transfect this latter human cell line. However, it is unlikely that transfection is a major factor in asbestos carcinogenesis – calcium apatite crystals are even more effective at transfection than asbestos fibres, and the former are present as part of tissue calcification whenever there is prolonged chronic inflammation; such inflammation is not regularly associated with carcinogenesis.

Fibres and DNA viruses
Much of the work on transformation or transfection has been undertaken using viral DNA and, in particular, using elements of the genome of the SV40 virus (Appel et al., 1988; Gan et al., 1993). This raises the possibility that, while transfection with exogenous DNA in general is not important in asbestos carcinogenesis, transfection of viral elements might increase the possibility of viral transformation of cells. SV40-like DNA has been identified in 60% of a series of human mesotheliomas (Carbone et al., 1994). This suggests that viral transformation with and without fibre–DNA transport may be important, and Ilgren and Wagner (1991) listed a series of incidences where virus infection had been associated with mesothelial tumour formation. One factor causing concern is that polio immunization in the 1950s involved injecting a live polio virus, some batches of which are now believed to have been contaminated with SV40 virus.

Fibres and radiation
One further factor that could act with asbestos or other mineral fibres in the causation of neoplasia is radiation. Certainly, there appears to be a synergistic effect between radon and cigarette smoking in the production of bronchial carcinomas (Lundin et al., 1969). Radiation has also been associated with the production of some mesotheliomas in humans (Ilgren & Wagner, 1991). Radon exposure in the general environment can be significant in some areas, and combined exposure to fibres and radiation is therefore quite likely. In addition, cigarette smoke contains significant amounts of polonium-210 nucleotide; smokers inhaling fibres have a triple exposure to fibre, PAH and radiation.

That there can be a synergistic effect between asbestos and radiation in tumour production is supported by some experimental evidence. Warren et al. (1981) examined the ability of injected chrysotile asbestos to induce pleural mesotheliomas in rats with and without concurrent X-ray irradiation. The production of mesotheliomas with chrysotile alone was only 3.8% (which is unusually low for a rat injection study), but this figure was raised to 6.1% in rats also receiving radiation. Similar results were obtained by Bignon et al. (1983) who injected rats with a variety of asbestos preparations (chrysotile, crocidolite, amosite) with subsequent irradiation with high LETα particles from radon gas. Since no rat received asbestos alone, the effect of radiation on mesothelioma production by asbestos could not be determined; however, radiation alone produced significant numbers of pulmonary carcinomas and the proportion of animals developing these tumours or mesotheliomas was increased by concurrent asbestos treatment. Hei et al. (1984) demonstrated that asbestos fibres could significantly increase the transformation rate of irradiated C3H 10T1/2 cells in vitro. While neither amosite or crocidolite fibres caused transformation when administered alone, they caused increased transformation of cells treated with γ-rays. This effect was not altered by the acid leaching of the asbestos fibres (Hei et al. 1985). Asbestos fibres combined with high energy α radiation produced a greater increase in cell transformation than a combination of asbestos and low energy γ radiation (Hei, 1989).

References
Appel, J.D., Fasy, T.M., Kohtz, D.S., Kohtz, J.D. & Johnson, E.M. (1988) Asbestos fibres mediate transformation of monkey cells by exogenous plasmid DNA. *Proc. Natl Acad. Sci. USA*, 85, 7670–7674

Avery, O.T., MacLeod, C.M. & McCarty, J. (1944) Studies of the chemical nature of the substance inducing tranformation of pneumococcal types. Induction of transformation by a deoxyribonucleic acid fraction isolated from Pneumococcus type III. *J. Exp. Med.*, 79, 137–158

Bignon, J., Monchaux, G., Chameaud, J., Jaurand, M.C., Lafuma, J. & Masse, R. (1983) Incidence of various types of thoracic malignancy induced in rats by the intrapleural injection of 2 mg of various mineral dusts after inhalation of ^{222}Ra. *Carcinogenesis*, 4, 621–628

Binns, R. (1975) Animal inhalation studies with tobacco smoke. *Rev. Environ. Health*, 2, 81–87

Bohning, D.E., Atkins, H.L. & Cohn, S.H. (1982) Long term particle clearance in man. Normal and impaired. *Ann. Occup. Hyg.*, 26, 259–271

Canmer, P. & Philipson, K. (1971) Intra-individual studies of tracheobronchial clearance in man using fluorocarbon resin particles tagged with ^{18}F and ^{44}m$_{Tc}$. In: Walton, W.H., ed., *Inhaled Particles III*, London, Unwin, pp. 157–180

Carbone, M., Pass, H.I., Rizzo, P., Marinetti, M.R., Dimuzio, M., Mew, D.J.Y., Levine, A.S. & Procopio, A. (1994) Simian virus 40-like DNA sequences in human pleural mesothelioma. *Oncogene*, 9, 1781–1790

Churg, A. & Wright, A.L. (1989) An animal model of co-exposure to cigarette smoke and mineral dusts. In: Wehner, A.P., ed., *Biological Interaction of Inhaled Mineral Fibers and Cigarette Smoke*, Columbus, OH, Battelle Press, pp. 85–96

Churg, A., Hobson, J. & Wright, J. (1989) Effects of cigarette smoke on uptake of asbestos fibres by trachael organ cultures: the role of active oxygen species. *Life Sci.*, 223, 447–455

Cohen, D., Arai, S.F. & Brain, J.D. (1979) Smoking impairs long term dust clearance from the lung. *Science*, 204, 514–417

Dagle, G.E., Wehner, A.P., Clark, M.L. & Buschbom, R.L. (1986) Chronic inhalation exposure of rats to quartz. In: Goldsmith, D.F., Winn, D.M. & Shy, C.H., eds, *Silica, Silicosis and Cancer*, New York, Praeger, pp. 225–266

Davis, J.M.G. (1989) Mineral fibre carcinogenesis: experimental data relating to the importance of fibre type, size, deposition, dissolution and migration. In: Bignon, J., Peto, J. & Saracci, Rd., eds, *Non-Occupational Exposure to Mineral Fibres* (IARC Scientific Publications No. 90), Lyon, IARC, pp. 33–45

Davis, J.M.G., Beckett, S.T., Bolton, R.E. & Donaldson, K. (1980) A comparison of the pathological effects in rats of the UICC reference samples of amosite and chysotile with those of amosite and chrysotile collected from the factory environment. In: Wagner, J.C., ed., *Biological Effects of Mineral Fibres* (IARC Scientific Publications No. 30), Lyon, IARC, pp. 285–292

Davis, J.M.G., Addison, J., Bolton, R.E., Donaldson, K., Jones, A.D. & Smith, T. (1986) The pathogenicity of long versus short fibre samples of amosite asbestos administered to rats by inhalation and intraperitoneal injection. *Br. J. Exp. Pathol.*, 67, 415–430

Davis, J.M.G. Addison, J., McIntosh, C., Miller, B.G. & Niven, K. (1991a) Variations in the carcinogenicity of tremolite dust samples of differing morphology. *Ann. N.Y. Acad. Sci.*, 643, 473–491

Davis, J.M.G., Jones, A.D. & Miller, B.G. (1991b) Experimental studies in rats on the effects of asbestos inhalation coupled with the inhalation of titanium dioxide or quartz. *Int. J. Exp. Pathol.*, 72, 501–525

Davis, J.M.G., Donaldson, K., Brown, D.M., Cullen, R.T., Jones, A.D., Miller, B.G., McIntosh, C., Searl, A. & Whittington, M. (1996) A comparison of methods of determining and predicting the pathogenicity of mineral fibres. *Inhal. Toxicol.* (in press)

Dubes, G.R. & Klinger, E.A. (1961) Facilitation of infection of monkey cells with polio virus ribonucleic acid. *Science*, 133, 99–100

Ferin, J.G., Urbankova, G. & Vickova, A. (1965) Influence of tobacco smoke on the elimination of particles from the lung. *Nature*, 206, 515–516

Finkelstein, M.M. (1983) Mortality among long term employees of an Ontario asbestos–cement factory. *Br. J. Ind. Med*, 40, 138–144

Gan, L., Savransky, E.F., Fasy, T.M. & Johnson, E.M. (1993) Transfection of human mesothelial cells medicated by different asbestos fiber types. *Environ. Res.*, 62, 28–42

Gerde, P., Muggenburg, B.A., Henderson, R.F. & Dahl, A.R. (1994) Particle-associated hydrocarbons and lung cancer. The correlation between cellular dosimetry and tumor distribution. *Cell Biol.*, 85, 337–344

Goodglick, L.A. & Kane, A.B. (1990) Cytotoxicity of long and short crocidollite asbestos fibers in vitro and in vivo. *Cancer Res.*, 50, 5153–5163

Graham, F.L. & van der Eb, A.J. (1973) A new technque for the assay of infectivity of human adenovirus 5 DNA. *Virology*, 52, 456–467

Harrington, J.S., Allison, A.C. & Badami, D.V. (1974) Mineral fibres: chemical, physicochemical and biological properties. *Adv. Pharmacol. Chemother.*, 12, 291–302

Harrison, P.T.C. & Heath, J.C. (1991) Apparent promotion by chrysotile asbestos of NHMI-initiated lung tumours in the rat. *Life Sci.*, 223, 469–479

Hei, T.K. (1989) Oncogenic transformation by asbestos fibers and radon-simulated alpha particles. *Cell Biol.*, 30, 389–397

Hei, T.K., Hall, E.J. & Osmak, R.S. (1984) Asbestos, radiation and oncogenic transformation. *Br. J. Cancer*, 50, 717–720

Hei, T.K., Geard, C.R., Osmak, R.S. & Travisano, M. (1985) Correlation of in-vitro genotoxicity and oncogenicity induced by radiation and asbestos fibers. *Br. J. Cancer*, 52, 591–597

Holland, L.-M., Wilson, J.S., Tillery, M.I. & Smith, D.M. (1986) Lung cancer in rats exposed to fibrogenic dusts. In: Goldsmith, D.F., Winn, D.M. &. Shy, C.M., eds, *Silica, Silicosis and Cancer*, New York, Praeger, pp. 267–280

Ilgren, E.B. & Wagner, J.C. (1991) Background incidence of mesothelioma: animal and human evidence. *Reg. Toxicol. Pharmacol.*, 13, 133–149

Jones, J.S.P. (1987) *Pathology of the Mesothelium*, New York, Springer-Verlag

Kandaswami, C. & O'Brien, P.J. (1980) Effects of asbestos on membrane transport and metabolism of benzo[a]pyrene. *Biochem. Biophys. Res. Commun.*, 97, 794–801

Lakowicz, J.R. & Bevan, D.R. (1980) Benzo[a]pyrene uptake into rat liver microsomes: effects of adsorption of benzo[a]pyrene to asbestos and non-fibrous mineral particles. *Chem. Biol. Interactions*, 29, 129–138

Lee, K.P., Trochimowicz, H.J. & Reinhardt, C.F. (1985) Pulmonary response of rats exposed to titanium dioxide (TiO_2) by inhalation for two years. *Toxicol. Appl. Pharmacol.*, 79, 179–192

Lourenco, R.V., Klimek, M.F. & Borowski, C.J. (1971) Deposition and clearance of 2 micron particles in the tracheobroncheal tree of normal subjects – smokers and nonsmokers. *J. Clin. Invest.*, 50, 1411–1419

Luchsinger, P.C., La Garde, B. & Kilfeather, J.E. (1968) Particle clearance from the human tracheobronchial tree. *Am. Rev. Respir. Dis.*, 97, 1046–1050

Luchtel, D.L., Martin, T.R. & Boatman, E.S. (1989) Characterisation of dusts from machining of fiber-epoxy composites and biological responses to respirable fractions. In: Wehner, A.P., ed., *Biological Interaction of Inhaled Mineral Fibers and Cigarette Smoke*, Columbus, OH, Battelle Press, pp. 291–313

Lundin, F.E., Lloyd, J.W., Smith, E.M., Archer, V.E. & Holaday, D.A. (1969) Mortality of uranium miners in relation to radiation exposure, hard rock mining and cigarette smoking. *Health Phys.*, 16, 571–578

Mauderley, J.L., Chen, B.T., Hahn, F.F., Lundgren, D.L., Cuddihy, R.G., Namenyi, J. & Rebar, A.H. (1989) The effect of chronic cigarette smoke inhalation on the long term pulmonary clearance of inhaled particles in the rat. In: Wehner, A.P., ed., *Biological Interaction of Inhaled Mineral Fibers and Cigarette Smoke*, Columbus, OH, Battelle Press, pp. 223–240

McFadden, J.L., Wright, L., Wiggs, B. & Churg, A. (1986) Smoking inhibits asbestos clearance. *Am. Rev. Respir. Dis.*, 133, 372–374

Muhle, H., Bellman, B., Spurney, K.R. & Pott, F. (1989) Inhalation experiments on retention and lung clearance of asbestos in combination with cigarette smoking. In: Wehner, A.P., ed., *Biological Interaction of Inhaled Mineral Fibers and Cigarette Smoke*, Columbus, OH, Battelle Press, pp. 183–194

Muhle, H., Bellmann, B., Greutzenberg, O., Dasenbrock, C., Ernst, H., Kilpper, R., MacKenzie, J.C., Morrow, P., Mohr, U., Takenaka, S. & Mermelstein, R. (1991) Pulmonary response to toner upon chronic inhalation in rats. *Fundam. Appl. Toxicol.*, 17, 280–299

Park, S.S., Kikkawa, Y., Golding, I.P., Daly, M.M., Zeletsky, M., Shim, C.H., Spierer, M. & Morita, T. (1977) An animal model of cigarette smoking in beagle dogs: correlative evaluation of effects on pulmonary function, defence and morphology. *Am. Rev. Respir. Dis.*, 115, 971–979

Poole, A., Brown, R.C. & Fleming, G.T.A. (1983) Study of the cell transforming ability of amosite and crocidolite asbestos and the ability to induce changes in the metabolism and macromolecular binding of benzo[a]pyrene in C3H10T cells. *Environ. Health Perspectives*, 51, 319–324

Pott, F. & Friedrichs, K.H. (1972) Tumoren der Ratte nach i.p. Injektion fraserformiger Staube. *Naturwissenschaften*, 59, 318

Pylev, L.N. & Shabad, L.M. (1973) Some results of experimental studies in asbestos carcinogenesis. In: Bogovski, P., Timbrel, U., Gilson, J.C. & Wagner, J.C., eds, *Biological Effects of Asbestos* (IARC Scientific Publications No. 8), Lyon, pp. 99–105

Reiss, B., Tong, C., Telang, S. & Williams, G.M. (1983) Enhancement of benzo[a]pyrene mutagenicity of chrysotile asbestos in rat liver epithelial cells. *Environ. Res.*, 31, 100–104

Rey, F., Boutin, C., Dumortier, P., Viallat, J.R. & De Vuyst, P. (1994) Carcinogenic asbestos fibers in the parietal pleura. *Cell Biol.*, 85, 311–317

Schepers, G.W.H. (1959a) Pulmonary histologic reactions to inhaled fiber-glass plastic dust. *Am. J. Pathol.*, 35, 1169–1189

Schepers, G.W.H. (1959b) The pulmonary reaction to sheet fiber glass plastic dust. *Am. Ind. Hyg. Assoc. J.*, 20, 73–81

Selikoff, I.J., Hammond, E.C. & Churg, J. (1968) Asbestos exposure, smoking and neoplasia. *J. Am. Med. Assoc.*, 204, 106–112

Shabad, L.M. Pylev, L.N., Krivosheeva, L.V., Kulagina, T.F. & Nemenko, B.A. (1974) Experimental studies on asbestos carcinogenicity. *J. Natl Cancer Inst.*, 52, 1175–1180

Stanton, M.F. & Wrench, C. (1972) Mechanisms of mesothelioma induction with asbestos and fibrous glass. *J. Natl Cancer Inst.*, 48, 797–821

Thomson, M.L. & Pavia, D. (1973) Long term tobacco smoking and mucociliary clearance. *Arch. Environ. Health*, 26, 88–91

Vincent, J.H., Jones, A.D., Johnston, A.M., McMillan, C., Bolton, R.E. & Cowie, H. (1987) Accumulation of inhaled mineral dust in the lung and associated lymph nodes: implications for exposure and dose in occupational lung disease. *Ann. Occup. Hyg.*, 31, 375–393

Warren, S., Brown, C.E., Chute, R.N. & Federman, N. (1981) Mesothelioma relative to asbestos, radiation and methylcholanthrene. *Arch. Pathol. Lab. Med.*, 105, 305–312

Wehner, A.P. (1980) Effect of inhaled asbestos, asbestos plus cigarette smoke, asbestos–cement and talc baby powder in hamsters. In: Wagner, J.C., ed., *Biological Effects of Mineral Fibres* (IARC Scientific Publications No. 30), Lyon, IARC, pp. 373–376

Wehner, A.P., Busch, R.H., Olson, R.J. & Craig, D.K. (1975) Chronic inhalation of asbestos and cigarette smoke by hamsters. *Environ. Res.*, 10, 368–383

Wehner, A.P., Dagle, G.E. & Cannon, W.C. (1978a) Development of an animal model, techniques, and an exposure system to study the effects of asbestos cement dust inhalation. *Environ. Res.*, 16, 393–407

Wehner, A.P., Dagle, G.E., Cannon, W.C. & Buschbom, R.L. (1978b) Asbestos cement dust inhalation by hamsters. *Environ. Res.*, 17, 367–389

Corresponding author
J.M. Davis
Institute of Occupational Medicine, Roxburgh Place, Edinburgh EH8 9SE, United Kingdom

IARC Monographs on the Evaluation of Carcinogenic Risks to Humans

Volume 1
Some Inorganic Substances, Chlorinated Hydrocarbons, Aromatic Amines, N-Nitroso Compounds, and Natural Products
1972; 184 pages; ISBN 92 832 1201 0
(out of print)

Volume 2
Some Inorganic and Organometallic Compounds
1973; 181 pages; ISBN 92 832 1202 9
(out of print)

Volume 3
Certain Polycyclic Aromatic Hydrocarbons and Heterocyclic Compounds
1973; 271 pages; ISBN 92 832 1203 7
(out of print)

Volume 4
Some Aromatic Amines, Hydrazine and Related Substances, N-Nitroso Compounds and Miscellaneous Alkylating Agents
1974; 286 pages; ISBN 92 832 1204 5

Volume 5
Some Organochlorine Pesticides
1974; 241 pages; ISBN 92 832 1205 3
(out of print)

Volume 6
Sex Hormones
1974; 243 pages; ISBN 92 832 1206 1
(out of print)

Volume 7
Some Anti-Thyroid and Related Substances, Nitrofurans and Industrial Chemicals
1974; 326 pages; ISBN 92 832 1207 X
(out of print)

Volume 8
Some Aromatic Azo Compounds
1975; 357 pages; ISBN 92 832 1208 8

Volume 9
Some Aziridines, N-, S- and O-Mustards and Selenium
1975; 268 pages; ISBN 92 832 1209 6

Volume 10
Some Naturally Occurring Substances
1976; 353 pages; ISBN 92 832 1210 X
(out of print)

Volume 11
Cadmium, Nickel, Some Epoxides, Miscellaneous Industrial Chemicals and General Considerations on Volatile Anaesthetics
1976; 306 pages; ISBN 92 832 1211 8
(out of print)

Volume 12
Some Carbamates, Thiocarbamates and Carbazides
1976; 282 pages; ISBN 92 832 1212 6

Volume 13
Some Miscellaneous Pharmaceutical Substances
1977; 255 pages; ISBN 92 832 1213 4

Volume 14
Asbestos
1977; 106 pages; ISBN 92 832 1214 2
(out of print)

Volume 15
Some Fumigants, the Herbicides 2,4-D and 2,4,5-T, Chlorinated Dibenzodioxins and Miscellaneous Industrial Chemicals
1977; 354 pages; ISBN 92 832 1215 0
(out of print)

Volume 16
Some Aromatic Amines and Related Nitro Compounds – Hair Dyes, Colouring Agents and Miscellaneous Industrial Chemicals
1978; 400 pages; ISBN 92 832 1216 9

Volume 17
Some N-Nitroso Compounds
1978; 365 pages; ISBN 92 832 1217 7

Volume 18
Polychlorinated Biphenyls and Polybrominated Biphenyls
1978; 140 pages; ISBN 92 832 1218 5

Volume 19
Some Monomers, Plastics and Synthetic Elastomers, and Acrolein
1979; 513 pages; ISBN 92 832 1219 3
(out of print)

Volume 20
Some Halogenated Hydrocarbons
1979; 609 pages; ISBN 92 832 1220 7
(out of print)

Volume 21
Sex Hormones (II)
1979; 583 pages; ISBN 92 832 1521 4

Volume 22
Some Non-Nutritive Sweetening Agents
1980; 208 pages; ISBN 92 832 1522 2

Volume 23
Some Metals and Metallic Compounds
1980; 438 pages; ISBN 92 832 1523 0
(out of print)

Volume 24
Some Pharmaceutical Drugs
1980; 337 pages; ISBN 92 832 1524 9

Volume 25
Wood, Leather and Some Associated Industries
1981; 412 pages; ISBN 92 832 1525 7

Volume 26
Some Antineoplastic and Immunosuppressive Agents
1981; 411 pages; ISBN 92 832 1526 5

Volume 27
Some Aromatic Amines, Anthraquinones and Nitroso Compounds, and Inorganic Fluorides Used in Drinking Water and Dental Preparations
1982; 341 pages; ISBN 92 832 1527 3

Volume 28
The Rubber Industry
1982; 486 pages; ISBN 92 832 1528 1

Volume 29
Some Industrial Chemicals and Dyestuffs
1982; 416 pages; ISBN 92 832 1529 X

Volume 30
Miscellaneous Pesticides
1983; 424 pages; ISBN 92 832 1530 3

Volume 31
Some Food Additives, Feed Additives and Naturally Occurring Substances
1983; 314 pages; ISBN 92 832 1531 1

Volume 32
Polynuclear Aromatic Compounds, Part 1: Chemical, Environmental and Experimental Data
1983; 477 pages; ISBN 92 832 1532 X

Volume 33
Polynuclear Aromatic Compounds, Part 2: Carbon Blacks, Mineral Oils and Some Nitroarenes
1984; 245 pages; ISBN 92 832 1533 8
(out of print)

Volume 34
Polynuclear Aromatic Compounds, Part 3: Industrial Exposures in Aluminium Production, Coal Gasification, Coke Production, and Iron and Steel Founding
1984; 219 pages; ISBN 92 832 1534 6

Volume 35
Polynuclear Aromatic Compounds: Part 4: Bitumens, Coal-Tars and Derived Products, Shale-Oils and Soots
1985; 271 pages; ISBN 92 832 1535 4

Volume 36
Allyl Compounds, Aldehydes, Epoxides and Peroxides
1985; 369 pages; ISBN 92 832 1536 2

Volume 37
Tobacco Habits Other than Smoking; Betel-Quid and Areca-Nut Chewing; and Some Related Nitrosamines
1985; 291 pages; ISBN 92 832 1537 0

Volume 38
Tobacco Smoking
1986; 421 pages; ISBN 92 832 1538 9

Volume 39
Some Chemicals Used in Plastics and Elastomers
1986; 403 pages; ISBN 92 832 1239 8

Volume 40
Some Naturally Occurring and Synthetic Food Components, Furocoumarins and Ultraviolet Radiation
1986; 444 pages; ISBN 92 832 1240 1

Volume 41
Some Halogenated Hydrocarbons and Pesticide Exposures
1986; 434 pages; ISBN 92 832 1241 X

Volume 42
Silica and Some Silicates
1987; 289 pages; ISBN 92 832 1242 8

Volume 43
Man-Made Mineral Fibres and Radon
1988; 300 pages; ISBN 92 832 1243 6

Volume 44
Alcohol Drinking
1988; 416 pages; ISBN 92 832 1244 4

Volume 45
Occupational Exposures in Petroleum Refining; Crude Oil and Major Petroleum Fuels
1989; 322 pages; ISBN 92 832 1245 2

Volume 46
Diesel and Gasoline Engine Exhausts and Some Nitroarenes
1989; 458 pages; ISBN 92 832 1246 0

Volume 47
Some Organic Solvents, Resin Monomers and Related Compounds, Pigments and Occupational Exposures in Paint Manufacture and Painting
1989; 535 pages; ISBN 92 832 1247 9

Volume 48
Some Flame Retardants and Textile Chemicals, and Exposures in the Textile Manufacturing Industry
1990; 345 pages; ISBN: 92 832 1248 7

Volume 49
Chromium, Nickel and Welding
1990; 677 pages; ISBN: 92 832 1249 5

Volume 50
Some Pharmaceutical Drugs
1990; 415 pages; ISBN: 92 832 1259 9

Volume 51
Coffee, Tea, Mate, Methylxanthines and Methylglyoxal
1991; 513 pages; ISBN: 92 832 1251 7

Volume 52
Chlorinated Drinking-Water; Chlorination By-products; Some other Halogenated Compounds; Cobalt and Cobalt Compounds
1991; 544 pages; ISBN: 92 832 1252 5

Volume 53
Occupational Exposures in Insecticide Application, and Some Pesticides
1991; 612 pages; ISBN 92 832 1253 3

Volume 54
Occupational Exposures to Mists and Vapours from Strong Inorganic Acids; and other Industrial Chemicals
1992; 336 pages; ISBN 92 832 1254 1

Volume 55
Solar and Ultraviolet Radiation
1992; 316 pages; ISBN 92 832 1255 X

Volume 56
Some Naturally Occurring Substances: Food Items and Constituents, Heterocyclic Aromatic Amines and Mycotoxins
1993; 600 pages; ISBN 92 832 1256 8

Volume 57
Occupational Exposures of Hairdressers and Barbers and Personal Use of Hair Colourants; Some Hair Dyes, Cosmetic Colourants, Industrial Dyestuffs and Aromatic Amines
1993; 428 pages; ISBN 92 832 1257 6

Volume 58
Beryllium, Cadmium, Mercury and Exposures in the Glass Manufacturing Industry
1994; 444 pages; ISBN 92 832 1258 4

Volume 59
Hepatitis Viruses
1994; 286 pages; ISBN 92 832 1259 2

Volume 60
Some Industrial Chemicals
1994; 560 pages; ISBN 92 832 1260 6

Volume 61
Schistosomes, Liver Flukes and *Helicobacter pylori*
1994; 280 pages; ISBN 92 832 1261 4

Volume 62
Wood Dusts and Formaldehyde
1995; 405 pages; ISBN 92 832 1262 2

Volume 63
Dry cleaning, Some Chlorinated Solvents and Other Industrial Chemicals
1995; 558 pages; ISBN 92 832 1263 0

Volume 64
Human Papillomaviruses
1995; 409 pages; ISBN 92 832 1264 9

Volume 65
Printing Processes, Printing Inks, Carbon Blacks and Some Nitro Compounds
1996; 578 pages; ISBN 92 832 1265 7

Volume 66
Some Pharmaceutical Drugs
1996; 514 pages; ISBN 92 832 1266 5

Supplements

Supplement No.1
Chemicals and Industrial Processes Associated with Cancer in Humans (IARC Monographs, Volumes 1 to 20)
1979; 71 pages; ISBN 92 832 1404 8
(out of print)

Supplement No. 2
Long-Term and Short-Term Screening Assays for Carcinogens: A Critical Appraisal
1980; 426 pages; ISBN 92 832 1404 8

Supplement No. 3
Cross Index of Synonyms and Trade Names in Volumes 1 to 26
1982; 199 pages; ISBN 92 832 1405 6
(out of print)

Supplement No.4
Chemicals, Industrial Processes and Industries Associated with Cancer in Humans (IARC Monographs, Volumes 1 to 29)
1982; 292 pages; ISBN 92 832 1407 2
(out of print)

Supplement No. 5
Cross Index of Synonyms and Trade Names in Volumes 1 to 36
1985; 259 pages; ISBN 92 832 1408 0
(out of print)

Supplement No. 6
Genetic and Related Effects: An Updating of Selected IARC Monographs from Volumes 1 to 42
1987; 729 pages; ISBN 92 832 1409 9

Supplement No. 7
Overall Evaluations of Carcinogenicity: An Updating of IARC Monographs Volumes 1 to 42
1987; 440 pages; ISBN 92 832 1411 0

Supplement No. 8
Cross Index of Synonyms and Trade Names in Volumes 1 to 46
1989; 346 pages; ISBN 92 832 1417 X

IARC Scientific Publications

No. 1
Liver Cancer
1971; 176 pages; ISBN 0 19 723000 8

No. 2
Oncogenesis and Herpesviruses
Edited by P.M. Biggs, G. de Thé and L.N. Payne
1972; 515 pages; ISBN 0 19 723001 6

No. 3
N-Nitroso Compounds: Analysis and Formation
Edited by P. Bogovski, R. Preussman and E.A. Walker
1972; 140 pages; ISBN 0 19 723002 4

No. 4
Transplacental Carcinogenesis
Edited by L. Tomatis and U. Mohr
1973; 181 pages; ISBN 0 19 723003 2

No. 5/6
Pathology of Tumours in Laboratory Animals. Volume 1: Tumours of the Rat
Edited by V.S. Turusov
1973/1976; 533 pages; ISBN 92 832 1410 2

No. 7
Host Environment Interactions in the Etiology of Cancer in Man
Edited by R. Doll and I. Vodopija
1973; 464 pages; ISBN 0 19 723006 7

No. 8
Biological Effects of Asbestos
Edited by P. Bogovski, J.C. Gilson, V. Timbrell and J.C. Wagner
1973; 346 pages; ISBN 0 19 723007 5

No. 9
N-Nitroso Compounds in the Environment
Edited by P. Bogovski and E.A. Walker
1974; 243 pages; ISBN 0 19 723008 3

No. 10
Chemical Carcinogenesis Essays
Edited by R. Montesano and L. Tomatis
1974; 230 pages; ISBN 0 19 723009 1

No. 11
Oncogenesis and Herpes-viruses II
Edited by G. de-Thé, M.A. Epstein and H. zur Hausen
1975; Two volumes, 511 pages and 403 pages; ISBN 0 19 723010 5

No. 12
Screening Tests in Chemical Carcinogenesis
Edited by R. Montesano, H. Bartsch and L. Tomatis
1976; 666 pages; ISBN 0 19 723051 2

No. 13
Environmental Pollution and Carcinogenic Risks
Edited by C. Rosenfeld and W. Davis
1975; 441 pages; ISBN 0 19 723012 1

No. 14
Environmental N-Nitroso Compounds. Analysis and Formation
Edited by E.A. Walker, P. Bogovski and L. Griciute
1976; 512 pages; ISBN 0 19 723013 X

No. 15
Cancer Incidence in Five Continents, Volume III
Edited by J.A.H. Waterhouse, C. Muir, P. Correa and J. Powell
1976; 584 pages; ISBN 0 19 723014 8

No. 16
Air Pollution and Cancer in Man
Edited by U. Mohr, D. Schmähl and L. Tomatis
1977; 328 pages; ISBN 0 19 723015 6

No. 17
Directory of On-Going Research in Cancer Epidemiology 1977
Edited by C.S. Muir and G. Wagner
1977; 599 pages; ISBN 92 832 1117 0
(out of print)

No. 18
Environmental Carcinogens. Selected Methods of Analysis. Volume 1: Analysis of Volatile Nitrosamines in Food
Editor-in-Chief: H. Egan
1978; 212 pages; ISBN 0 19 723017 2

No. 19
Environmental Aspects of N-Nitroso Compounds
Edited by E.A. Walker, M. Castegnaro, L. Griciute and R.E. Lyle
1978; 561 pages; ISBN 0 19 723018 0

No. 20
Nasopharyngeal Carcinoma: Etiology and Control
Edited by G. de Thé and Y. Ito
1978; 606 pages; ISBN 0 19 723019 9

No. 21
Cancer Registration and its Techniques
Edited by R. MacLennan, C. Muir, R. Steinitz and A. Winkler
1978; 235 pages; ISBN 0 19 723020 2

No. 22
Environmental Carcinogens: Selected Methods of Analysis. Volume 2: Methods for the Measurement of Vinyl Chloride in Poly(vinyl chloride), Air, Water and Foodstuffs
Editor-in-Chief: H. Egan
1978; 142 pages; ISBN 0 19 723021 0

No. 23
Pathology of Tumours in Laboratory Animals. Volume II: Tumours of the Mouse
Editor-in-Chief: V.S. Turusov
1979; 669 pages; ISBN 0 19 723022 9

No. 24
Oncogenesis and Herpesviruses III
Edited by G. de-Thé, W. Henle and F. Rapp
1978; Part I: 580 pages, Part II: 512 pages; ISBN 0 19 723023 7

No. 25
Carcinogenic Risk: Strategies for Intervention
Edited by W. Davis and C. Rosenfeld
1979; 280 pages; ISBN 0 19 723025 3

No. 26
Directory of On-going Research in Cancer Epidemiology 1978
Edited by C.S. Muir and G. Wagner
1978; 550 pages; ISBN 0 19 723026 1
(out of print)

No. 27
Molecular and Cellular Aspects of Carcinogen Screening Tests
Edited by R. Montesano, H. Bartsch and L. Tomatis
1980; 372 pages; ISBN 0 19 723027 X

No. 28
Directory of On-going Research in Cancer Epidemiology 1979
Edited by C.S. Muir and G. Wagner
1979; 672 pages; ISBN 92 832 1128 6
(out of print)

No. 29
Environmental Carcinogens. Selected Methods of Analysis. Volume 3: Analysis of Polycyclic Aromatic Hydrocarbons in Environmental Samples
Editor-in-Chief: H. Egan
1979; 240 pages; ISBN 0 19 723028 8

No. 30
Biological Effects of Mineral Fibres
Editor-in-Chief: J.C. Wagner
1980; Two volumes, 494 pages & 513 pages; ISBN 0 19 723030 X

No. 31
N-Nitroso Compounds: Analysis, Formation and Occurrence
Edited by E.A. Walker, L. Griciute, M. Castegnaro and M. Börzsönyi
1980; 835 pages; ISBN 0 19 723031 8

No. 32
Statistical Methods in Cancer Research. Volume 1: The Analysis of Case-control Studies
By N.E. Breslow and N.E. Day
1980; 338 pages; ISBN 92 832 0132 9

No. 33
Handling Chemical Carcinogens in the Laboratory
Edited by R. Montesano, H. Bartsch, E. Boyland, G. Della Porta, L. Fishbein, R.A. Griesemer, A.B. Swan and L. Tomatis
1979; 32 pages; ISBN 0 19 723033 4
(out of print)

No. 34
Pathology of Tumours in Laboratory Animals. Volume III: Tumours of the Hamster
Editor-in-Chief: V.S. Turusov
1982; 461 pages; ISBN 0 19 723034 2

No. 35
Directory of On-going Research in Cancer Epidemiology 1980
Edited by C.S. Muir and G. Wagner
1980; 660 pages; ISBN 0 19 723035 0
(out of print)

No. 36
Cancer Mortality by Occupation and Social Class 1851–1971
Edited by W.P.D. Logan
1982; 253 pages; ISBN 0 19 723036 9

No. 37
Laboratory Decontamination and Destruction of Aflatoxins B1, B2, G1, G2 in Laboratory Wastes
Edited by M. Castegnaro, D.C. Hunt, E.B. Sansone, P.L. Schuller, M.G. Siriwardana, G.M. Telling, H.P. van Egmond and E.A. Walker
1980; 56 pages; ISBN 0 19 723037 7

No. 38
Directory of On-going Research in Cancer Epidemiology 1981
Edited by C.S. Muir and G. Wagner
1981; 696 pages; ISBN 0 19 723038 5
(out of print)

No. 39
Host Factors in Human Carcinogenesis
Edited by H. Bartsch and B. Armstrong
1982; 583 pages;
ISBN 0 19 723039 3

No. 40
Environmental Carcinogens: Selected Methods of Analysis. Volume 4: Some Aromatic Amines and Azo Dyes in the General and Industrial Environment
Edited by L. Fishbein, M. Castegnaro, I.K. O'Neill and H. Bartsch
1981; 347 pages; ISBN 0 19 723040 7

No. 41
N-Nitroso Compounds: Occurrence and Biological Effects
Edited by H. Bartsch, I.K. O'Neill, M. Castegnaro and M. Okada
982; 755 pages; ISBN 0 19 723041 5

No. 42
Cancer Incidence in Five Continents Volume IV
Edited by J. Waterhouse, C. Muir, K. Shanmugaratnam and J. Powell
1982; 811 pages; ISBN 0 19 723042 3

No. 43
Laboratory Decontamination and Destruction of Carcinogens in Laboratory Wastes: Some N-Nitrosamines
Edited by M. Castegnaro, G. Eisenbrand, G. Ellen, L. Keefer, D. Klein, E.B. Sansone, D. Spincer, G. Telling and K. Webb
1982; 73 pages; ISBN 0 19 723043 1

No. 44
Environmental Carcinogens: Selected Methods of Analysis.
Volume 5: Some Mycotoxins
Edited by L. Stoloff, M. Castegnaro, P. Scott, I.K. O'Neill and H. Bartsch
1983; 455 pages; ISBN 0 19 723044 X

No. 45
Environmental Carcinogens: Selected Methods of Analysis.
Volume 6: N-Nitroso Compounds
Edited by R. Preussmann, I.K. O'Neill, G. Eisenbrand, B. Spiegelhalder and H. Bartsch
1983; 508 pages; ISBN 0 19 723045 8

No. 46
Directory of On-going Research in Cancer Epidemiology 1982
Edited by C.S. Muir and G. Wagner
1982; 722 pages; ISBN 0 19 723046 6
(out of print)

No. 47
Cancer Incidence in Singapore 1968–1977
Edited by K. Shanmugaratnam, H.P. Lee and N.E. Day
1983; 171 pages; ISBN 0 19 723047 4

No. 48
Cancer Incidence in the USSR (2nd Revised Edition)
Edited by N.P. Napalkov, G.F. Tserkovny, V.M. Merabishvili, D.M. Parkin, M. Smans and C.S. Muir
1983; 75 pages; ISBN 0 19 723048 2

No. 49
Laboratory Decontamination and Destruction of Carcinogens in Laboratory Wastes: Some Polycyclic Aromatic Hydrocarbons
Edited by M. Castegnaro, G. Grimmer, O. Hutzinger, W. Karcher, H. Kunte, M. Lafontaine, H.C. Van der Plas, E.B. Sansone and S.P. Tucker
1983; 87 pages; ISBN 0 19 723049 0

No. 50
Directory of On-going Research in Cancer Epidemiology 1983
Edited by C.S. Muir and G. Wagner
1983; 731 pages; ISBN 0 19 723050 4
(out of print)

No. 51
Modulators of Experimental Carcinogenesis
Edited by V. Turusov and R. Montesano
1983; 307 pages; ISBN 0 19 723060 1

No. 52
Second Cancers in Relation to Radiation Treatment for Cervical Cancer: Results

of a Cancer Registry Collaboration
Edited by N.E. Day and J.C. Boice, Jr
1984; 207 pages; ISBN 0 19 723052 0

No. 53
Nickel in the Human Environment
Editor-in-Chief: F.W. Sunderman, Jr
1984; 529 pages; ISBN 0 19 723059 8

No. 54
Laboratory Decontamination and Destruction of Carcinogens in Laboratory Wastes: Some Hydrazines
Edited by M. Castegnaro, G. Ellen, M. Lafontaine, H.C. van der Plas, E.B. Sansone and S.P. Tucker
1983; 87 pages; ISBN 0 19 723053

No. 55
Laboratory Decontamination and Destruction of Carcinogens in Laboratory Wastes: Some N-Nitrosamides
Edited by M. Castegnaro, M. Bernard, L.W. van Broekhoven, D. Fine, R. Massey, E.B. Sansone, P.L.R. Smith, B. Spiegelhalder, A. Stacchini, G. Telling and J.J. Vallon
1984; 66 pages; ISBN 0 19 723054 7

No. 56
Models, Mechanisms and Etiology of Tumour Promotion
Edited by M. Börzsönyi, N.E. Day, K. Lapis and H. Yamasaki
1984; 532 pages; ISBN 0 19 723058 X

No. 57
N-Nitroso Compounds: Occurrence, Biological Effects and Relevance to Human Cancer
Edited by I.K. O'Neill, R.C. von Borstel, C.T. Miller, J. Long and H. Bartsch
1984; 1013 pages; ISBN 0 19 723055 5

No 58
Age-related Factors in Carcinogenesis
Edited by A. Likhachev, V. Anisimov and R. Montesano
1985; 288 pages; ISBN 92 832 1158 8

No. 59
Monitoring Human Exposure to Carcinogenic and Mutagenic Agents
Edited by A. Berlin, M. Draper, K. Hemminki and H. Vainio
1984; 457 pages; ISBN 0 19 723056 3

No. 60
Burkitt's Lymphoma: A Human Cancer Model
Edited by G. Lenoir, G. O'Conor and C.L.M. Olweny
1985; 484 pages; ISBN 0 19 723057 1

No. 61
Laboratory Decontamination and Destruction of Carcinogens in Laboratory Wastes: Some Haloethers
Edited by M. Castegnaro, M. Alvarez, M. Iovu, E.B. Sansone, G.M. Telling and D.T. Williams
1985; 55 pages; ISBN 0 19 723061 X

No. 62
Directory of On-going Research in Cancer Epidemiology 1984
Edited by C.S. Muir and G. Wagner
1984; 717 pages; ISBN 0 19 723062 8
(out of print)

No. 63
Virus-associated Cancers in Africa
Edited by A.O. Williams, G.T. O'Conor, G.B. de Thé and C.A. Johnson
1984; 773 pages; ISBN 0 19 723063 6

No. 64
Laboratory Decontamination and Destruction of Carcinogens in Laboratory Wastes: Some Aromatic Amines and 4-Nitrobiphenyl
Edited by M. Castegnaro, J. Barek, J. Dennis, G. Ellen, M. Klibanov, M. Lafontaine, R. Mitchum, P. van Roosmalen, E.B. Sansone, L.A. Sternson and M. Vahl
1985; 84 pages; ISBN: 92 832 1164 2

No. 65
Interpretation of Negative Epidemiological Evidence for Carcinogenicity
Edited by N.J. Wald and R. Doll
1985; 232 pages; ISBN 92 832 1165 0

No. 66
The Role of the Registry in Cancer Control
Edited by D.M. Parkin, G. Wagner and C.S. Muir
1985; 152 pages; ISBN 92 832 0166 3

No. 67
Transformation Assay of Established Cell Lines: Mechanisms and Application
Edited by T. Kakunaga and H. Yamasaki
1985; 225 pages; ISBN 92 832 1167 7

No. 68
Environmental Carcinogens: Selected Methods of Analysis. Volume 7: Some Volatile Halogenated Hydrocarbons
Edited by L. Fishbein and I.K. O'Neill
1985; 479 pages; ISBN 92 832 1168 5

No. 69
Directory of On-going Research in Cancer Epidemiology 1985
Edited by C.S. Muir and G. Wagner
1985; 745 pages; ISBN 92 823 1169 3
(out of print)

No. 70
The Role of Cyclic Nucleic Acid Adducts in Carcinogenesis and Mutagenesis
Edited by B. Singer and H. Bartsch
1986; 467 pages; ISBN 92 832 1170 7

No. 71
Environmental Carcinogens: Selected Methods of Analysis. Volume 8: Some Metals: As, Be, Cd, Cr, Ni, Pb, Se, Zn
Edited by I.K. O'Neill, P. Schuller and L. Fishbein
1986; 485 pages; ISBN 92 832 1171 5

No. 72
Atlas of Cancer in Scotland, 1975–1980: Incidence and Epidemiological Perspective
Edited by I. Kemp, P. Boyle, M. Smans and C.S. Muir
1985; 285 pages; ISBN 92 832 1172 3

No. 73
Laboratory Decontamination and Destruction of Carcinogens in Laboratory Wastes: Some Antineoplastic Agents
Edited by M. Castegnaro, J. Adams, M.A. Armour, J. Barek, J. Benvenuto, C. Confalonieri, U. Goff, G. Telling
1985; 163 pages; ISBN 92 832 1173 1

No. 74
Tobacco: A Major International Health Hazard
Edited by D. Zaridze and R. Peto
1986; 324 pages; ISBN 92 832 1174 X

No. 75
Cancer Occurrence in Developing Countries
Edited by D.M. Parkin
1986; 339 pages; ISBN 92 832 1175 8

No. 76
Screening for Cancer of the Uterine Cervix
Edited by M. Hakama, A.B. Miller and N.E. Day
1986; 315 pages; ISBN 92 832 1176 6

No. 77
Hexachlorobenzene: Proceedings of an International Symposium
Edited by C.R. Morris and J.R.P. Cabral
1986; 668 pages; ISBN 92 832 1177 4

No. 78
Carcinogenicity of Alkylating Cytostatic Drugs
Edited by D. Schmähl and J.M. Kaldor
1986; 337 pages; ISBN 92 832 1178 2

No. 79
Statistical Methods in Cancer Research. Volume III: The Design and Analysis of Long-term Animal Experiments
By J.J. Gart, D. Krewski, P.N. Lee, R.E. Tarone and J. Wahrendorf
1986; 213 pages; ISBN 92 832 1179 0

No. 80
Directory of On-going Research in Cancer Epidemiology 1986
Edited by C.S. Muir and G. Wagner
1986; 805 pages; ISBN 92 832 1180 4
(out of print)

No. 81
Environmental Carcinogens: Methods of Analysis and Exposure Measurement. Volume 9: Passive Smoking
Edited by I.K. O'Neill, K.D. Brunnemann, B. Dodet and D. Hoffmann
1987; 383 pages; ISBN 92 832 1181 2

No. 82
Statistical Methods in Cancer Research. Volume II: The Design and Analysis of Cohort Studies
By N.E. Breslow and N.E. Day
1987; 404 pages; ISBN 92 832 0182 5

No. 83
Long-term and Short-term Assays for Carcinogens: A Critical Appraisal
Edited by R. Montesano, H. Bartsch, H. Vainio, J. Wilbourn and H. Yamasaki
1986; 575 pages; ISBN 92 832 1183 9

No. 84
The Relevance of N-Nitroso Compounds to Human Cancer: Exposure and Mechanisms
Edited by H. Bartsch, I.K. O'Neill and R. Schulte-Hermann
1987; 671 pages; ISBN 92 832 1184 7

No. 85
Environmental Carcinogens: Methods of Analysis and Exposure Measurement. Volume 10: Benzene and Alkylated Benzenes
Edited by L. Fishbein and I.K. O'Neill
1988; 327 pages; ISBN 92 832 1185 5

No. 86
Directory of On-going Research in Cancer Epidemiology 1987
Edited by D.M. Parkin and J. Wahrendorf
1987; 685 pages; ISBN: 92 832 1186 3
(out of print)

No. 87
International Incidence of Childhood Cancer
Edited by D.M. Parkin, C.A. Stiller, C.A. Bieber, G.J. Draper. B. Terracini and J.L. Young
1988; 401 page; ISBN 92 832 1187 1
(out of print)

No. 88
Cancer Incidence in Five Continents, Volume V
Edited by C. Muir, J. Waterhouse, T. Mack, J. Powell and S. Whelan
1987; 1004 pages; ISBN 92 832 1188 X

No. 89
Methods for Detecting DNA Damaging Agents in Humans: Applications in Cancer Epidemiology and Prevention
Edited by H. Bartsch, K. Hemminki and I.K. O'Neill
1988; 518 pages; ISBN 92 832 1189 8
(out of print)

No. 90
Non-occupational Exposure to Mineral Fibres
Edited by J. Bignon, J. Peto and R. Saracci
1989; 500 pages; ISBN 92 832 1190 1

No. 91
Trends in Cancer Incidence in Singapore 1968–1982
Edited by H.P. Lee, N.E. Day and K. Shanmugaratnam
1988; 160 pages; ISBN 92 832 1191 X

No. 92
Cell Differentiation, Genes and Cancer
Edited by T. Kakunaga, T. Sugimura, L. Tomatis and H. Yamasaki
1988; 204 pages; ISBN 92 832 1192 8

No. 93
Directory of On-going Research in Cancer Epidemiology 1988
Edited by M. Coleman and J. Wahrendorf
1988; 662 pages; ISBN 92 832 1193 6
(out of print)

No. 94
Human Papillomavirus and Cervical Cancer
Edited by N. Muñoz, F.X. Bosch and O.M. Jensen
1989; 154 pages; ISBN 92 832 1194 4

No. 95
Cancer Registration: Principles and Methods
Edited by O.M. Jensen, D.M. Parkin, R. MacLennan, C.S. Muir and R. Skeet
1991; 296 pages; ISBN 92 832 1195 2

No. 96
Perinatal and Multigeneration Carcinogenesis
Edited by N.P. Napalkov, J.M. Rice, L. Tomatis and H. Yamasaki
1989; 436 pages; ISBN 92 832 1196 0

No. 97
Occupational Exposure to Silica and Cancer Risk
Edited by L. Simonato, A.C. Fletcher, R. Saracci and T. Thomas
1990; 124 pages; ISBN 92 832 1197 9

No. 98
Cancer Incidence in Jewish Migrants to Israel, 1961-1981
Edited by R. Steinitz, D.M. Parkin, J.L. Young, C.A. Bieber and L. Katz
1989; 320 pages; ISBN 92 832 1198 7

No. 99
Pathology of Tumours in Laboratory Animals, Second Edition, Volume 1, Tumours of the Rat
Edited by V.S. Turusov and U. Mohr
1990; 740 pages; ISBN 92 832 1199 5
For Volumes 2 and 3 (Tumours of the Mouse and Tumours of the Hamster), see IARC Scientific Publications Nos. 111 and 126.

No. 100
Cancer: Causes, Occurrence and Control
Editor-in-Chief: L. Tomatis
1990; 352 pages; ISBN 92 832 0110 8

No. 101
Directory of On-going Research in Cancer Epidemiology 1989–1990
Edited by M. Coleman and J. Wahrendorf
1989; 828 pages; ISBN 92 832 2101 X

No. 102
Patterns of Cancer in Five Continents
Edited by S.L. Whelan, D.M. Parkin and E. Masuyer
1990; 160 pages; ISBN 92 832 2102 8

No. 103
Evaluating Effectiveness of Primary Prevention of Cancer
Edited by M. Hakama, V. Beral, J.W. Cullen and D.M. Parkin
1990; 206 pages; ISBN 92 832 2103 6

No. 104
Complex Mixtures and Cancer Risk
Edited by H. Vainio, M. Sorsa and A.J. McMichael
1990; 441 pages; ISBN 92 832 2104 4

No. 105
Relevance to Human Cancer of N-Nitroso Compounds, Tobacco Smoke and Mycotoxins
Edited by I.K. O'Neill, J. Chen and H. Bartsch
1991; 614 pages; ISBN 92 832 2105 2

No. 106
Atlas of Cancer Incidence in the Former German Democratic Republic
Edited by W.H. Mehnert, M. Smans, C.S. Muir, M. Möhner and D. Schön
1992; 384 pages; ISBN 92 832 2106 0

No. 107
Atlas of Cancer Mortality in the European Economic Community
Edited by M. Smans, C. Muir and P. Boyle
1992; 213 pages + 44 coloured maps; ISBN 92 832 2107 9

No. 108
Environmental Carcinogens: Methods of Analysis and Exposure Measurement. Volume 11: Polychlorinated Dioxins and Dibenzofurans
Edited by C. Rappe, H.R. Buser, B. Dodet and I.K. O'Neill
1991; 400 pages; ISBN 92 832 2108 7

No. 109
Environmental Carcinogens: Methods of Analysis and Exposure Measurement. Volume 12: Indoor Air
Edited by B. Seifert, H. van de Wiel, B. Dodet and I.K. O'Neill
1993; 385 pages; ISBN 92 832 2109 5

No. 110
Directory of On-going Research in Cancer Epidemiology 1991
Edited by M.P. Coleman and J. Wahrendorf
1991; 753 pages; ISBN 92 832 2110 9

No. 111
Pathology of Tumours in Laboratory Animals, Second Edition. Volume 2: Tumours of the Mouse
Edited by V. Turusov and U. Mohr
1994; 800 pages; ISBN 92 832 2111 1

No. 112
Autopsy in Epidemiology and Medical Research
Edited by E. Riboli and M. Delendi
1991; 288 pages; ISBN 92 832 2112 5

No. 113
Laboratory Decontamination and Destruction of Carcinogens in Laboratory Wastes: Some Mycotoxins
Edited by M. Castegnaro, J. Barek, J.M. Frémy, M. Lafontaine, M. Miraglia, E.B. Sansone and G.M. Telling
1991; 63 pages; ISBN 92 832 2113 3

No. 114
Laboratory Decontamination and Destruction of Carcinogens in Laboratory Wastes: Some Polycyclic Heterocyclic Hydrocarbons
Edited by M. Castegnaro, J. Barek, J. Jacob, U. Kirso, M. Lafontaine, E.B. Sansone, G.M. Telling and T. Vu Duc
1991; 50 pages; ISBN 92 832 2114 1

No. 115
Mycotoxins, Endemic Nephropathy and Urinary Tract Tumours
Edited by M. Castegnaro, R. Plestina, G. Dirheimer, I.N. Chernozemsky and H. Bartsch
1991; 340 pages; ISBN 92 832 2115 X

No. 116
Mechanisms of Carcinogenesis in Risk Identification
Edited by H. Vainio, P. Magee, D. McGregor and A.J. McMichael
1992; 615 pages; ISBN 92 832 2116 8

No. 117
Directory of On-going Research in Cancer Epidemiology 1992
Edited by M. Coleman, E. Demaret and J. Wahrendorf
1992; 773 pages; ISBN 92 832 2117 6

No. 118
Cadmium in the Human Environment: Toxicity and Carcinogenicity
Edited by G.F. Nordberg, R.F.M. Herber and L. Alessio
1992; 470 pages; ISBN 92 832 2118 4

No. 119
The Epidemiology of Cervical Cancer and Human Papillomavirus
Edited by N. Muñoz, F.X. Bosch, K.V. Shah and A. Meheus
1992; 288 pages; ISBN 92 832 2119 2

No. 120
Cancer Incidence in Five Continents, Vol. VI
Edited by D.M. Parkin, C.S. Muir, S.L. Whelan, Y.T. Gao, J. Ferlay and J. Powell
1992; 1020 pages; ISBN 92 832 2120 6

No. 121
Time Trends in Cancer Incidence and Mortality
By M. Coleman, J. Estéve, P. Damiecki, A. Arslan and H. Renard
1993; 820 pages; ISBN 92 832 2121 4

No. 122
International Classification of Rodent Tumours.
Part I. The Rat
Editor-in-Chief: U. Mohr
1992–1996; 10 fascicles of 60–100 pages; ISBN 92 832 2122 2

No. 123
Cancer in Italian Migrant Populations
Edited by M. Geddes, D.M. Parkin, M. Khlat, D. Balzi and E. Buiatti
1993; 292 pages; ISBN 92 832 2123 0

No. 124
Postlabelling Methods for the Detection of DNA Damage
Edited by D.H. Phillips, M. Castegnaro and H. Bartsch
1993; 392 pages; ISBN 92 832 2124 9

No. 125
DNA Adducts: Identification and Biological Significance
Edited by K. Hemminki, A. Dipple, D.E.G. Shuker, F.F. Kadlubar, D. Segerbäck and H. Bartsch
1994; 478 pages; ISBN 92 832 2125 7

No. 126
Pathology of Tumours in Laboratory Animals, Second Edition. Volume 3: Tumours of the Hamster
Edited by V. Turosov and U. Mohr
1996; 464 pages; ISBN 92 832 2126 5

No. 128
Statistical Methods in Cancer Research. Volume IV. Descriptive Epidemiology
By J. Estève, E. Benhamou and L. Raymond
1994; 302 pages; ISBN 92 832 2128 1

No. 129
Occupational Cancer in Developing Countries
Edited by N. Pearce, E. Matos, H. Vainio, P. Boffetta and M. Kogevinas
1994; 191 pages; ISBN 92 832 2129 X

No. 130
Directory of On-going Research in Cancer Epidemiology 1994
Edited by R. Sankaranarayanan, J. Wahrendorf and E. Démaret
1994; 800 pages; ISBN 92 832 2130 3

No. 132
Survival of Cancer Patients in Europe: The EUROCARE Study
Edited by F. Berrino, M. Sant, A. Verdecchia, R. Capocaccia, T. Hakulinen and J. Estève
1995; 463 pages; ISBN 92 832 2132 X

No. 134
Atlas of Cancer Mortality in Central Europe
W. Zatonski, J. Estéve, M. Smans, J. Tyczynski and P. Boyle
1996; 300 pages; ISBN 92 832 2134 6

No. 136
Chemoprevention in Cancer Control
Edited by M. Hakama, V. Beral, E. Buiatti, J. Faivre and D.M. Parkin
1996; 160 pages; ISBN 92 832 2136 2

No. 137
Directory of On-going Research in Cancer Epidemiology 1996
Edited by R. Sankaranarayan, J. Warendorf and E. Démaret
1996; 810 pages; ISBN 92 832 2137 0

No. 139
Principles of Chemoprevention
Edited by B.W. Stewart, D. McGregor and P. Kleihues
1996; 360 pages;
ISBN 92 832 2139 7

No. 140
Mechanisms of Fibre Carcinogenesis
Edited by A.B. Kane, P. Boffetta, R. Saracci and J.D. Wilbourn
1996; 135 pages; ISBN 92 832 2140 0

IARC Technical Reports

No. 1
Cancer in Costa Rica
Edited by R. Sierra, R. Barrantes, G. Muñoz Leiva, D.M. Parkin, C.A. Bieber and N. Muñoz Calero
1988; 124 pages; ISBN 92 832 1412 9

No. 2
SEARCH: A Computer Package to Assist the Statistical Analysis of Case-Control Studies
Edited by G.J. Macfarlane, P. Boyle and P. Maisonneuve
1991; 80 pages; ISBN 92 832 1413 7

No. 3
Cancer Registration in the European Economic Community
Edited by M.P. Coleman and E. Démaret
1988; 188 pages; ISBN 92 832 1414 5

No. 4
Diet, Hormones and Cancer: Methodological Issues for Prospective Studies
Edited by E. Riboli and R. Saracci
1988; 156 pages; ISBN 92 832 1415 3

No. 5
Cancer in the Philippines
Edited by A.V. Laudico, D. Esteban and D.M. Parkin
1989; 186 pages; ISBN 92 832 1416 1

No. 6
La genèse du Centre international de recherche sur le cancer
By R. Sohier and A.G.B. Sutherland
1990, 102 pages; ISBN 92 832 1418 8

No. 7
Epidémiologie du cancer dans les pays de langue latine
1990, 292 pages; ISBN 92 832 1419 6

No. 8
Comparative Study of Anti-smoking Legislation in Countries of the European Economic Community
By A. J. Sasco, P. Dalla-Vorgia and P. Van der Elst
1992; 82 pages; ISBN: 92 832 1421 8
Etude comparative des Législations de Contrôle du Tabagisme dans les Pays de la Communauté économique européenne
1995; 82 pages; ISBN 92 832 2402 7

No. 9
Epidémiologie du cancer dans les pays de langue latine
1991; 346 pages; ISBN 92 832 1423 4

No. 10
Manual for Cancer Registry Personnel
Edited by D. Esteban, S. Whelan, A. Laudico and D.M. Parkin
1995; 400 pages; ISBN 92 832 1424 2

No. 11
Nitroso Compounds: Biological Mechanisms, Exposures and Cancer Etiology
Edited by I. O'Neill and H. Bartsch
1992; 150 pages; ISBN 92 832 1425 X

No. 12
Epidémiologie du cancer dans les pays de langue latine
1992; 375 pages; ISBN 92 832 1426 9

No. 13
Health, Solar UV Radiation and Environmental Change
By A. Kricker, B.K. Armstrong, M.E. Jones and R.C. Burton
1993; 213 pages; ISBN 92 832 1427 7

No. 14
Epidémiologie du cancer dans les pays de langue latine
1993; 400 pages; ISBN 92 832 1428 5

No. 15
Cancer in the African Population of Bulawayo, Zimbabwe, 1963–1977
By M.E.G. Skinner, D.M. Parkin, A.P. Vizcaino and A. Ndhlovu
1993; 120 pages; ISBN 92 832 1429 3

No. 16
Cancer in Thailand 1984–1991
By V. Vatanasapt, N. Martin, H. Sriplung, K. Chindavijak, S. Sontipong, S. Sriamporn, D.M. Parkin and J. Ferlay
1993; 164 pages; ISBN 92 832 1430 7

No. 18
Intervention Trials for Cancer Prevention
By E. Buiatti
1994; 52 pages; ISBN 92 832 1432 3

No. 19
Comparability and Quality Control in Cancer Registration
By D.M. Parkin, V.W. Chen, J. Ferlay, J. Galceran, H.H. Storm and S.L. Whelan
1994; 110 pages plus diskette;
ISBN 92 832 1433 1

No. 20
Epidémiologie du cancer dans les pays de langue latine
1994; 346 pages; ISBN 92 832 1434 X

No. 21
ICD Conversion Programs for Cancer
By J. Ferlay
1994; 24 pages plus diskette;
ISBN 92 832 1435 8

No. 22
Cancer in Tianjin
By Q.S. Wang, P. Boffetta, M. Kogevinas and D.M. Parkin
1994; 96 pages; ISBN 92 832 1433 1

No. 23
An Evaluation Programme for Cancer Preventive Agents
By Bernard W. Stewart
1995; 40 pages; ISBN 92 832 1438 2

No. 24
Peroxisome Proliferation and its Role in Carcinogenesis
1995; 85 pages; ISBN 92 832 1439 0

No. 25
Combined Analysis of Cancer Mortality in Nuclear Workers in Canada, the United Kingdom and the United States of America
By E. Cardis, E.S. Gilbert, L. Carpenter, G. Howe, I. Kato, J. Fix, L. Salmon, G. Cowper, B.K. Armstrong, V. Beral, A. Douglas, S.A. Fry, J. Kaldor, C. Lavé, P.G. Smith, G. Voelz and L. Wiggs
1995; 160 pages; ISBN 92 832 1440 4

Directories of Agents being Tested for Carcinogenicity
Edited by M.-J. Ghess, J.D. Wilbourn and H. Vainio

No. 15
1992; 317 pages; ISBN 92 832 1315 7

No. 16
1994; 294 pages; ISBN 92 832 1316 5

No. 17
1996; 360 pages; ISBN 92 832 1317 3

Non-serial publications
Alcool et Cancer
By A. Tuyns
1978; 48 pages

Cancer Morbidity and Causes of Death among Danish Brewery Workers
By O.M. Jensen
1980; 143 pages

Directory of Computer Systems Used in Cancer Registries
By H.R. Menck and D.M. Parkin
1986; 236 pages

Facts and Figures of Cancer in the European Community
By J. Estève, A. Kricker, J. Ferlay and D.M. Parkin
1993; 52 pages; ISBN 92 832 1437 4

All IARC Publications are available directly from
IARCPress, 150 Cours Albert Thomas, F-69372 Lyon cedex 08, France
(Fax: +33 4 72 73 83 02; E-mail: press@iarc.fr).

IARC Monographs and Technical Reports are also available from the
World Health Organization Distribution and Sales, CH-1211 Geneva 27
(Fax: +41 22 791 4857)
and from WHO Sales Agents worldwide.

IARC Scientific Publications are also available from
Oxford University Press, Walton Street, Oxford, UK OX2 6DP
(Fax: +44 1865 267782).